T0140175

Springer Proceedings in Mathematics & Statistics

Volume 301

Springer Proceedings in Mathematics & Statistics

This book series features volumes composed of selected contributions from workshops and conferences in all areas of current research in mathematics and statistics, including operation research and optimization. In addition to an overall evaluation of the interest, scientific quality, and timeliness of each proposal at the hands of the publisher, individual contributions are all refereed to the high quality standards of leading journals in the field. Thus, this series provides the research community with well-edited, authoritative reports on developments in the most exciting areas of mathematical and statistical research today.

More information about this series at http://www.springer.com/series/10533

Isadora Antoniano-Villalobos ·
Ramsés H. Mena · Manuel Mendoza ·
Lizbeth Naranjo · Luis E. Nieto-Barajas
Editors

Selected Contributions on Statistics and Data Science in Latin America

33 FNE and 13 CLATSE, 2018, Guadalajara, Mexico, October 1–5

Editors
Isadora Antoniano-Villalobos
Department of Environmental Sciences, Informatics and Statistics
Ca' Foscari University of Venice
Venice, Italy

Ramsés H. Mena
Department of Probability and Statistics
IIMAS, UNAM
Mexico City, Mexico

Manuel Mendoza
Department of Statistics
Instituto Tecnológico Autónomo de México
Mexico City, Mexico

Lizbeth Naranjo
Department of Mathematics
Facultad de Ciencias, UNAM
Mexico City, Mexico

Luis E. Nieto-Barajas
Department of Statistics
Instituto Tecnológico Autónomo de México
Mexico City, Mexico

ISSN 2194-1009 ISSN 2194-1017 (electronic)
Springer Proceedings in Mathematics & Statistics
ISBN 978-3-030-31553-5 ISBN 978-3-030-31551-1 (eBook)
https://doi.org/10.1007/978-3-030-31551-1

Mathematics Subject Classification (2010): 34K12, 34K29, 60J22, 60J25, 60J75, 62D05, 62F15, 62G99, 62H20, 62H25, 62H30, 62H99, 62J12, 62J99, 62M05, 62M09, 62P10, 65C40, 90C90, 92D30

This Springer imprint is published by the registered company Springer Nature Switzerland AG
The registered company address is: Gewerbestrasse 11, 6330 Cham, Switzerland

Preface

This volume includes a selection of peer-reviewed contributions presented at the 33FNE and 13CLATSE meetings held jointly in Guadalajara, Mexico from October 1st to 5th, 2018.

The FNE (Foro Nacional de Estadística) is the official meeting of the Mexican Statistical Association (AME), taking place annually since 1986. The purpose of the FNE is to offer an opportunity for statisticians and practitioners to share the latest developments in research and applications, exchange ideas, and explore opportunities for collaboration. The meetings are complemented by short courses and workshops, to fulfill the mission of AME of promoting the knowledge and good practice of Statistics in the country. The CLATSE (Congreso Latino Americano de Sociedades de Estadística) is the joint statistical meeting of Latin America. Born as a collaboration between the Argentinean (SAE) and Chilean (SOCHE) Statistical Societies, it has grown significantly since its first edition in 1991 in Valparaíso, Chile. It now includes the Statistical Societies of Colombia (SCE), Ecuador (SEE), Perú (SOPEST), and Uruguay (SUE), as well as AME and the Brazilian Statistical Association (ABE). The 33FNE was organized by AME and the University of Guadalajara (UDG), while responsibility for the 13CLATSE was shared by SAE, ABE, SOCHE, SCE, and AME. The joint event was hosted by the UDG University Centre for Exact Sciences and Ingeneering (CUCEI).

Statistical research in Latin America is prolific, and collaborative networks span within and outside the region. A great territorial extension, climatic peculiarities, and political and socioeconomic factors may hinder the international dissemination of the high-quality research output of the region. Additionally, much of the work is typically carried out and published in Spanish, and thus a large portion of the interested public may overlook interesting findings. We hope that this volume will provide access to selected works from Latin American statisticians and their research networks to a wider audience. We are sure that new methodological advances, motivated in part by the challenges of a data-driven world and the Latin American context, will be of interest to academics and practitioners around the world.

The scientific program of the 33FNE and 13CLATSE included a total of 107 oral presentations, organized in 43 contributed and 7 invited sessions, in Spanish and English, plus 4 keynote sessions delivered by Alexandra M. Schmidt (McGill University, Canada), Haavard Rue (KAUST, Saudi Arabia), Abel Rodriguez (University of California Santa Cruz, USA), and Francisco Louzada (University of São Paulo, Brazil). Five short courses on Spacial Statistics (Ronny Vallejos, Federico Santa María Technical University, Chile), Computational methods for Bayesian inference (Hedibert Lopes, Insper, Brazil), Environmental Statistics (Bruno Sansó, University of Santa Cruz, USA), The challenges of Teaching Statistics: New Scenarios at the Undergraduate, Masters and Doctorate levels (María Purifcación Galindo, University of Salamanca, Spain), Statistical Foundations of Machine Learning with STATA (Miguel Ángel Cruz, MultiON Consulting, Mexico), and 45 poster presentations completed the program. The event was preceded by a full day of courses, aimed mainly at interested students, on the topics of Variational Bayes and beyond: Bayesian inference for big data (Tamara Broderick, MIT, USA), Machine Learning (Elmer Garduño, Google Inc., USA), Bayesian computing with INLA (Haavard Rue, KAUST, Saudi Arabia), and Statistical and psychometric intricacies of educational survey assessments (Andreas Oranje, Educational Testing Service, USA).

We thank all participants who brought scientific quality to the events and made the experience rewarding. A special recognition is due to the local organizers Humberto Gutiérrez Pulido and Abelardo Montesinos López (UDG, Mexico), and to Leticia Ramírez Ramírez (CIMAT, Mexico) of the organizing committee. The international quality of the event would not have been achieved without the hard work of the members of the Scientific Committees: Eduardo Gutiérrez Peña (UNAM, Mexico), Abelardo Montesinos López (UDG, Mexico), Lizbeth Naranjo Albarrán (UNAM, Mexico), and Luis Enrique Nieto Barajas (ITAM, Mexico), for the 33FNE; Jorge Luis Bazán (University of São Paulo, Brazil), Ramón Giraldo (National University of Colombia), Manuel Mendoza (ITAM, Mexico), Orietta Nicolis (University of Valparaiso, Chile), and Lila Ricci (National University of Mar del Plata, Argentina), for the 13CLATSE.

Venice, Italy — Isadora Antoniano-Villalobos
Mexico City, Mexico — Ramsés H. Mena
Mexico City, Mexico — Manuel Mendoza
Mexico City, Mexico — Lizbeth Naranjo
Mexico City, Mexico — Luis E. Nieto-Barajas

Contents

A Heavy-Tailed Multilevel Mixture Model for the Quick Count in the Mexican Elections of 2018

Michelle Anzarut, Luis Felipe González and María Teresa Ortiz

Abstract Quick counts based on probabilistic samples are powerful methods for monitoring election processes. However, the complete designed samples are rarely collected to publish the results in a timely manner. Hence, the results are announced using partial samples, which have biases associated to the arrival pattern of the information. In this paper, we present a Bayesian hierarchical model to produce estimates for the Mexican gubernatorial elections. The model considers the poll stations poststratified by demographic, geographic, and other covariates. As a result, it provides a principled means of controlling for biases associated to such covariates. We compare methods through simulation exercises and apply our proposal in the July 2018 elections for governor in certain states. Our studies find the proposal to be more robust than the classical ratio estimator and other estimators that have been used for this purpose.

Keywords Bayesian calibration · Hierarchical model · Model-based inference · Multilevel regression · Poststratification · Zero-inflated model

1 Introduction

In this paper, we present one of the statistical models used in the quick count of the 2018 Mexican elections. Mexico is a Federal State that comprises 32 states. The government system is presidential; the president and the governor of each state are elected for a 6-year term by the population. The candidate who wins a plurality of

M. Anzarut (✉) · L. F. González · M. T. Ortiz
Instituto Tecnológico Autónomo de México, Río Hondo 1, Altavista 01080, CDMX, Mexico
e-mail: michelle@sigma.iimas.unam.mx

L. F. González
e-mail: luis.gonzalez@itam.mx

M. T. Ortiz
e-mail: maria.ortiz@itam.mx

I. Antoniano-Villalobos et al. (eds.), *Selected Contributions on Statistics and Data Science in Latin America*, Springer Proceedings in Mathematics & Statistics 301,
https://doi.org/10.1007/978-3-030-31551-1_1

votes is elected and no president nor governor may be reelected. Each state has its own electoral calendar, and in some cases, the federal and state elections coincide.

The National Electoral Institute (INE) is a public, autonomous agency with the authority for organizing elections. The INE organizes a quick count the same night of the election. The quick count consists of selecting a random sample of the polling stations and estimating the percentage of votes in favor of each candidate. With highly competed electoral processes, the rapidity and precision of the quick count results have become very important. Even more, the election official results are presented to the population a week after the election day. Therefore, the quick count prevents unjustified victory claims during that period.

The election of 2018 was qualified as the largest election that has taken place in Mexico, with 3,400 positions in dispute. For the first time, quick counts were made for nine local elections for the governor position, simultaneous to a quick count for the presidential federal election. The INE creates a committee of specialists in charge of the quick count, whose responsibilities encompass, mostly, the sample design, and the operation of statistical methods to produce the inferences. The inferences are presented as probability intervals with an associated probability of at least 0.95.

The information system starts at 6 p.m. and, every 5 min, collects all the sample information sent. Thus, the system produces a sequence of accumulative files used to determine the available percentage of the sample and its distribution over the country. The partial samples are analyzed with the estimation methods to track the trend of the results. Notice that the partial samples have a potential bias associated to the arrival pattern of the information. Generally, the quick count results that are made public use one of these partial samples, since the complete sample takes too long to arrive. The committee reports a result when certain conditions are met, such as the arrival of a large part of the sample and the stability in the estimates.

In addition to the partial samples being biased, it has been observed in recent elections that the complete planned sample hardly ever arrives. Studying the missing data of the 2012 elections, we note that the two main reasons for this missingness are communication problems in nonurban areas and the weather conditions in certain regions, especially heavy rain. Therefore, we must assume that the data is not missing completely at random. As a consequence, the probability that a polling station is not reported may depend on the response we intend to measure.

The context of the analysis is then as follows. We have a stratified sample designed by the committee, so we know the inclusion probabilities and the strata weights, which are proportional to the inverse of the inclusion probabilities. The key challenge is that we have to estimate with incomplete and biased samples, which may imply limited (or null) sample size in some strata and where the missingness is not completely at random. This is where model-based inference brings out its usefulness.

For a population with N units, let $Y = (y_1, ..., y_N)$ be the survey variables and $I = (I_1, ..., I_N)$ be the inclusion indicator variables, where $I_i = 1$ if unit i is included in the sample and $I_i = 0$ if it is not included. Design-based inference for a finite population quantity $f(Y)$ involves the choice of an estimator $\hat{f}(Y)$. The estimator is a function of the sampled values Y_{inc}, and usually is unbiased for f with respect to I.

Hence, the distribution of I remains the basis for inference. As an implication, design-based methods do not provide a consistent treatment when there is nonresponse or response errors. Model-based inference means modeling both Y and I. The model is used to predict the non-sampled values of the population, and hence finite population quantities. A more detailed explanation can be found in [6].

Let Z denote known design variables; we apply Bayesian model-based inference, meaning that we specify a prior distribution $P(Y|Z)$ for the population values. With this, we have the posterior predictive distribution $p(Y_{\text{exc}}|Y_{\text{inc}}, Z, I)$ where Y_{exc} are the non-sampled values. We make the inference for the total number of votes in favor of each candidate based on this posterior predictive distribution. Occasionally, it is possible to ignore the data collection mechanism. In such a case, the mechanism is called ignorable (see [4, p. 202]). This means that inferences are based on the posterior predictive distribution $p(Y_{\text{exc}}|Y_{\text{inc}}, Z)$, which simplifies the modeling task. However, it also means we are assuming that, conditional on Z, the missing data pattern supplies no information. With a complete sample, the mechanism would be ignorable by including the strata variable in the model. In this setting, there is always missing data. Hence, we include in the analysis all the explanatory variables we consider relevant in the data collection mechanism or in the vote decision. Note that as more explanatory variables are included, the ignorability assumption becomes more plausible.

The proposed model is a *heavy-tailed multilevel mixture model* (denoted as heavy-MM model). This model, defined later on, is a Bayesian multilevel regression, where the dependent variable has a heavy-tailed distribution with a mass at zero. We tested the heavy-MM model using data from 2006 and 2012 gubernatorial elections in the states of Chiapas, Morelos, and Guanajuato. The model was used, among others, to estimate the results in the quick count of the 2018 elections in those three states. In this paper, to show the process of model building, we use the data of Guanajuato.

The outline of the paper is as follows. In Sect. 2, we describe the sample design. In Sect. 3, we define the proposed model. In Sect. 4, we describe the estimation method and calibration. In Sect. 5, we develop the application of the model to the elections of 2018. Finally, in Sect. 6, we give some concluding remarks and future research directions.

2 Sample Design

The sample design was stratified where, within each stratum, we selected polling stations by simple random sampling without replacement. To define the strata, we considered possible combinations of the following variables:

- Federal district: Units in which the territory of Mexico is divided for the purpose of elections.
- Local district: Units in which the territory of each state is divided for the purpose of elections.

- Section type: Urban, rural, or mixed.

In addition, as a comparative point, we also considered simple random sampling without stratification.

For each combination, we computed the estimation precision with a 95% of probability and with different sample sizes using the databases of gubernatorial elections of 2012. The details may be consulted in [1]. The more variables used in the stratification, the smaller the estimation error. The same applies to the sample size, the greater the sample the smaller the error. Nevertheless, there are some other important criteria that need to be evaluated, for example, the total number of strata, the average number of polling stations within each stratum, and the number of strata with few voting stations. Moreover, we also took into consideration the average number of polling stations to be reported by field employees, and the percentage of field employees in charge of more than one polling station. The aim of evaluating all of these criteria is to find a balance that minimizes the errors without jeopardizing the collection of the sample.

After considering all the alternatives, we decided to use the local district with a sample of 500 units, giving rise to 22 strata with an average of 300 polling stations each. Finally, we set the sample size for each stratum proportionally to its size.

3 A Multilevel Model

Multilevel models are appropriate for research designs where data are nested. The model we propose is based on the well-known multilevel regression and poststratification model, which has a long history (see, for example, [10]), but its modern-day implementation can be traced to [9]. The central idea of the multilevel regression and poststratification model is to use multilevel regression to model individual survey responses as a function of different attributes, and then weight the estimates to estimate at the population level. The multilevel regression requires a set of predictors and the choice of a probability distribution. In this section, we discuss those two topics.

3.1 Predictors

For the elections, the INE does a geographic subdivision of the country in electoral sections. The electoral sections can be labeled as urban, rural, or mixed. Within each section, a basic polling station is installed. Additionally, other types of polling station may be installed which are:

- Adjoint polling station: They are installed when the number of voters in the section is greater than 750.

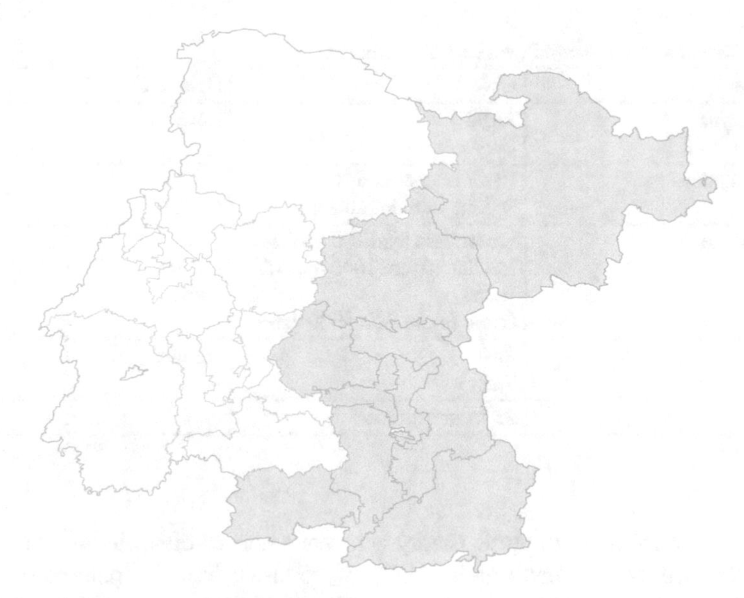

Fig. 1 State of Guanajuato divided by local district; the gray and white indicate the two regions considered as predictors in the heavy-MM model

- Extraordinary polling station: They are for the residents of a section that, because of conditions of communication routes or sociocultural matters, have difficult access to the polling stations.
- Special polling station: They are installed so that the voters outside the section corresponding to their home can vote.

As a consequence, at most 750 citizens are registered as potential voters on every station. The file with the names and photographs of these citizens is called the nominal list.

After testing all the available predictors, Table 1 summarizes the ones that we choose to use. The specification of the region variable can be found in Fig. 1. In addition, it is natural to consider the interaction of section type with section size. We model all the variables in Table 1, except strata, as regression coefficients without multilevel structure.

While exit polls and past election results could be strong predictors, we cannot include them in the model since it is considered to be politically unacceptable for a quick count organized by the electoral authority.

3.2 *Multilevel Model with Normal Probability Distribution*

Once we have established the predictors, we move to the task of defining the probability distribution assumed for the total number of votes of each candidate. A common

Table 1 Predictors of the multilevel regression model

Predictor	Levels	Notation
Section type	Rural Urban or mixed	Rural ·
Polling station type	Basic or contiguous Special or extraordinary	· typeSP
Section size	Small (less than 1000 voters) Medium (from 1000 to 5000 voters) Large (more than 5000 voters)	· sizeM sizeL
Region	East West	regionE ·
Strata	22 local districts	·

assumption based on asymptotic theory is to use a normal distribution. One contribution that follows this direction is [7]. They propose a Bayesian parametric model where the number of people in favor of a candidate divided by poll and stratum has a normal distribution. Based on the same idea, the first model we raise is the multilevel model with normal probability distribution.

We model each candidate independently, let X_k be the number of votes in favor of a candidate in the k-th polling station, and then

$$X_k \sim \mathsf{N}(\mu_k, \sigma_k^2)\, \mathsf{I}_{[0,750]}, \tag{1}$$

with mean $\mu_k = n_k\theta_k$ and variance $\sigma_k^2 = n_k\psi^2_{\text{strata}(k)}$. The indicator function $\mathsf{I}_{[0,750]}$ is one if the value is in the interval [0, 750] or zero otherwise. The term n_k is the size of the nominal list in the polling station, θ_k represents the proportion of people in the nominal list of the k-th polling station who voted for the candidate, and the variance $\psi^2_{\text{strata}(k)}$ is assumed to be constant in the corresponding stratum.

We fit a multilevel regression model for the parameter θ_k,

$$\begin{aligned}\theta_k = \text{logit}^{-1}(\beta^0 &+ \beta^{\text{rural}} \cdot \text{rural}_k + \beta^{\text{rural\_sizeM}} \cdot \text{rural}_k \cdot \text{sizeM}_k \\ &+\beta^{\text{sizeM}} \cdot \text{sizeM}_k + \beta^{\text{sizeL}} \cdot \text{sizeL}_k + \beta^{\text{regionE}} \cdot \text{regionE}_k \\ &+\beta^{\text{strata}}_{\text{strata}(k)} + \beta^{\text{typeSP}} \cdot \text{typeSP}_k).\end{aligned}$$

Finally, we adjust a model to the stratum level,

$$\beta_j^{\text{strata}} \sim \mathsf{N}\left(\mu^{\text{strata}}, \sigma^2_{\text{strata}}\right),$$

Where μ^{strata} is given a $\mathsf{N}(0, 10)$ initial distribution, and σ^2_{strata} is given a $\mathsf{U}(0, 5)$ initial distribution. For the rest of the coefficients, we also assign a $\mathsf{N}(0, 10)$ initial distribution.

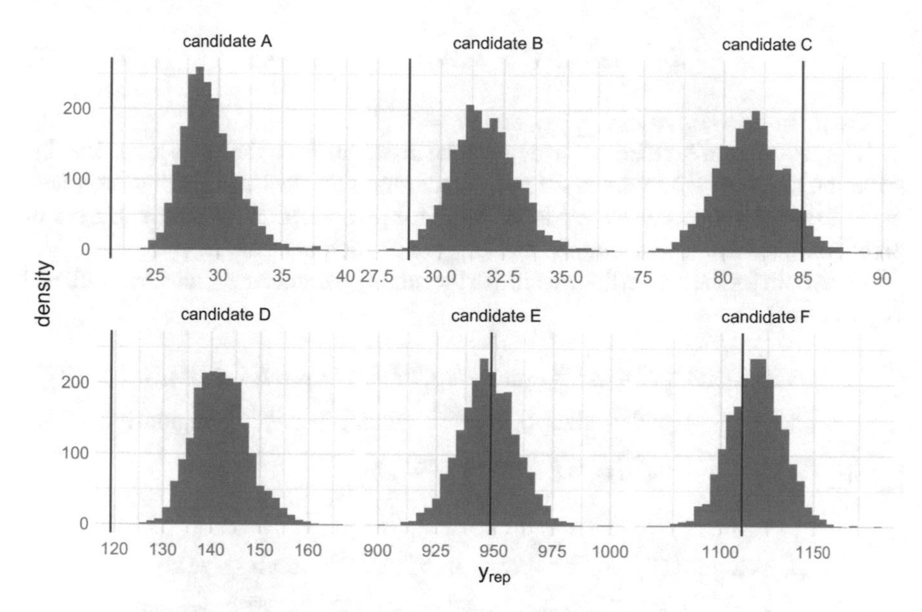

Fig. 2 Posterior predictive distributions of the total number of votes (in thousands) in the 2012 gubernatorial elections of Guanajuato using the multilevel model with normal probability distribution. The red line indicates the total number of votes observed

By adding predictors at the stratum level, we reduce the unexplained variation within each stratum and, as a result, we also reduce the total variation, producing more precise estimates.

As a first step to evaluate the model, we perform a posterior predictive check using the 2012 data. This check helps us test the richness of the model to capture the relevant structure of the true data generating process (see [2]). Figure 2 shows the posterior predictive distributions of the total number of votes. Clearly, the truncated normal distribution produces a bad fit. This is in part due to the longer tails of observed data compared to the normal distribution, in particular with smaller candidates, which tend to have districts where they are extremely popular compared to the rest of the state. Therefore, we need to use another type of probability distribution.

3.3 *The Heavy-MM Model*

We need a distribution that is also bell-shaped, but with heavier tails than the normal distribution. A natural choice is the t-distribution. However, we also need to catch the high number of zero votes in some polling stations. This leads us to the *heavy-MM model*, which is a multilevel model with a zero-inflated probability distribution. In this model, we replace distribution in Eq. (1) for

$$X_k \sim p_k\delta_0 + (1 - p_k)\mathsf{t}(\mu_k, \sigma_k^2, \nu_k)\, \mathsf{I}_{[0,750]}, \tag{2}$$

where $\mu_k = n_k\theta_k, \sigma_k^2 = n_k\psi^2_{\text{strata}(k)}$, and $\nu_k = \nu_k^{\text{district}}$.

The distribution δ_0 refers to the degenerate distribution on the value zero. The distribution $\mathsf{t}(\mu_k, \sigma_k^2, \nu_k)$ refers to the non-standardized Student's t-distribution, where μ_k is the location parameter, σ_k^2 is the variance parameter and ν_k is the degrees of freedom. The terms $\mathsf{I}_{[0,750]}$, n_k, θ_k, and $\psi^2_{\text{strata}(k)}$ are defined as in Eq. (1).

Now, both for the proportion θ_k, and for the mixing parameter p_k, we fit a multilevel regression

$$\begin{aligned}\theta_k = \text{logit}^{-1}(&\beta^0 + \beta^{\text{rural}} \cdot \text{rural}_k + \beta^{\text{rural\_sizeM}} \cdot \text{rural}_k \cdot \text{sizeM}_k \\ &+ \beta^{\text{sizeM}} \cdot \text{sizeM}_k + \beta^{\text{sizeL}} \cdot \text{sizeL}_k + \beta^{\text{regionE}} \cdot \text{regionE}_k \\ &+ \beta^{\text{strata}}_{\text{strata}(k)} + \beta^{\text{typeSP}} \cdot \text{typeSP}_k),\end{aligned}$$

$$\begin{aligned}p_k = \text{logit}^{-1}(&\beta^0_p + \beta^{\text{rural}}_p \cdot \text{rural}_k + \beta^{\text{rural\_sizeM}}_p \cdot \text{rural}_k \cdot \text{sizeM}_k \\ &+ \beta^{\text{sizeM}}_p \cdot \text{sizeM}_k + \beta^{\text{sizeL}}_p \cdot \text{sizeL}_k + \beta^{\text{regionE}}_p \cdot \text{regionE}_k \\ &+ \beta^{\text{strata-p}}_{\text{strata}(k)} + \beta^{\text{typeSP}}_p \cdot \text{typeSP}_k).\end{aligned}$$

Finally,

$$\begin{aligned}\beta_j^{\text{strata}} &\sim \mathsf{N}\left(\mu^{\text{strata}}, \sigma^2_{\text{strata}}\right), \\ \beta_j^{\text{strata-p}} &\sim \mathsf{N}\left(\mu^{\text{strata-p}}, \sigma^2_{\text{strata-p}}\right),\end{aligned}$$

Where μ^{strata} is given a $\mathsf{N}(0, 10)$ initial distribution, and σ^2_{strata} is given a $\mathsf{U}(0, 5)$ initial distribution. For the rest of the coefficients, we assign independent $\mathsf{N}(0, 10)$ initial distributions. The corresponding posterior distributions have unknown forms. Hence, they are approximated with simulation methods.

For the parameters $\beta^{\text{strata}}_{\text{strata}(k)}$ and $\beta^{\text{strata-p}}_{\text{strata}(k)}$, we use redundant parameters to speed computation, rescaling based on the mean (see [3, p. 316]). This can be consulted in detail in the R package `quickcountmx` [8].

The heavy-MM model assumes that the sample is a mixture of two sorts of individuals: one group whose counts are generated by the Student's t-distribution, and another group who has zero probability of a count greater than zero. This implies that the heavy-MM model is part of the zero-inflated models (see, for example, [5]), which have become fairly popular in the research literature.

Performing the same posterior predictive check as for the previous model, we can see in Fig. 3 the distribution (2) gives a much more appropriate fit. Therefore, we used the heavy-MM model to predict the results of the 2018 elections.

Notice that party symmetry is preserved in the model since no ad hoc adjustments are made for particular parties, their size, or their results in previous elections. The

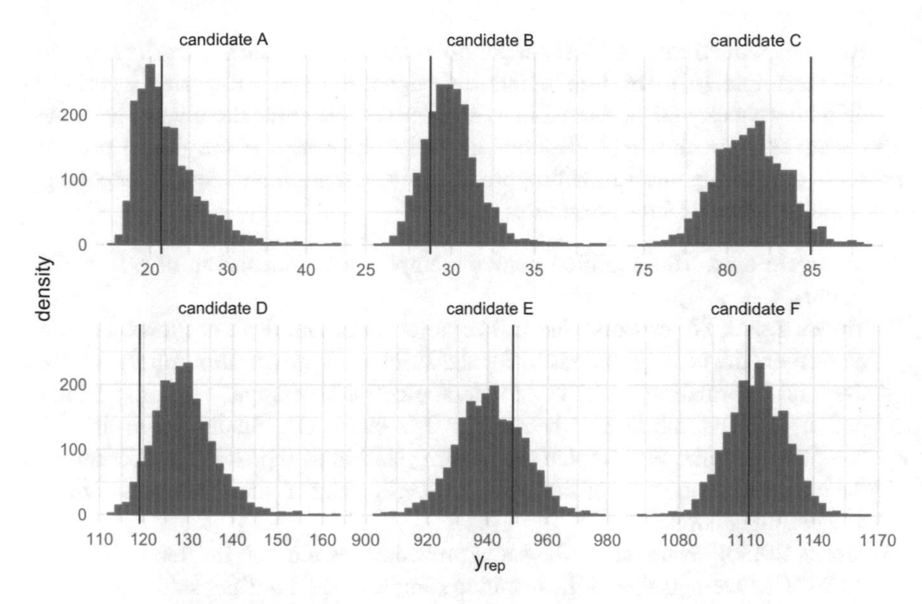

Fig. 3 Posterior predictive distributions of the total number of votes (in thousands) in the 2012 gubernatorial elections of Guanajuato using the heavy-MM model. The red line indicates the total number of votes observed

political scenario of Mexico has changed drastically in recent electoral periods, so this is an essential feature of the model.

4 Estimation and Calibration

We implemented the model using the software JAGS called from R. The implementation can be consulted and fully reproduced in the R package `quickcountmx` [8]. Let us stress out the importance of this reproducibility. On the scientific side, reproducibility is crucial to examine the veracity and robustness of the conclusions of a paper. However, in this case, reproducibility is more important still, as it helps to achieve transparency in the electoral procedure. Any citizen can download the sample and compute the same results that were announced publically the night of the election. This transparency fosters trust in institutions and gives legitimacy to the outcome of the quick count.

After fitting the model, the straightforward estimation proceeded as follows. First, for every polling station not in the sample, we simulated its vote counts according to the model. Then, we aggregated observed values from polling stations with simulated ones and obtained simulated vote counts for the total of polling stations. Finally, we summarized the aggregated samples to produce vote proportions and the corresponding intervals.

To ensure we can cover a variety of possible outcomes, we did Bayesian calibration (see, for example, [6]). This implies that, although the model is Bayesian, it is chosen to yield inferences well calibrated in a frequentist sense. Specifically, we generated 100 samples from the 2012 elections database and tested if the model provides posterior probability intervals with approximately their nominal frequentist coverage. We simulated five different scenarios:

1. Complete data: 100 stratified random samples each consisting of 507 polling stations.
2. Trends 22:00: We censored the 100 complete data samples as follows. First, we partitioned the polling stations by local district and type (rural/no-rural). We then used the information of the 2012 federal election to compute the proportion of polling stations that arrived before 22:00 in each cell. Finally, for each of the complete samples, we sampled each polling station with probability according to the observed proportion of the cell. The average size of the resulting samples is 445 polling stations.
3. Trends 20:30: We repeated the same procedure as the one in Trends 22:00 for 20:30. The average size of the resulting samples is 262 polling stations.
4. Strata biased: We censored the 100 complete data samples deleting all the polling stations from three strata. We selected the three strata independently for each sample, with probability proportional to the observed proportion of votes in the strata for the major candidate. The resulting samples average 438 polling stations.
5. Polls biased: We deleted 15% of the polling stations of each of the 100 complete data samples. We considered the probability of not observing a polling station proportional to the observed proportion of votes in the strata for the major candidate. The final sample size is of 431 polling stations.

We also did the calibration for two other methods. One is the ratio estimator, which is a traditional survey sampling estimation method. The other one is the Bayesian parametric model found in [7], which we refer to as the normal no pooling model (NNP). Our implementation of this model can be found in the `quickcountmx` package. These methods were chosen because versions of them were also used for the quick count during the 2018 elections.

We want methods producing well-calibrated estimates of the uncertainty. Also, intervals should be reasonably narrow to be useful in most situations (for example, within one percentage point of actual tallies). Additionally, we analyze the performance of each method. Estimation procedures should be fast enough to produce results as the batches of data are received so that partial samples can be monitored, and the decision of when to publish taken. In this case, the committee established the model fitting process should not take more than 10 min.

Figure 4 shows the calibration. Let us point out that all the results resemble each other when the estimation is done with the complete sample, and neither the ratio estimator nor the Bayesian method of [7] is designed to be used when there are missing polls or missing strata. However, on the night of the election, we estimate using partial samples. With either of these two methods, modifications can be made,

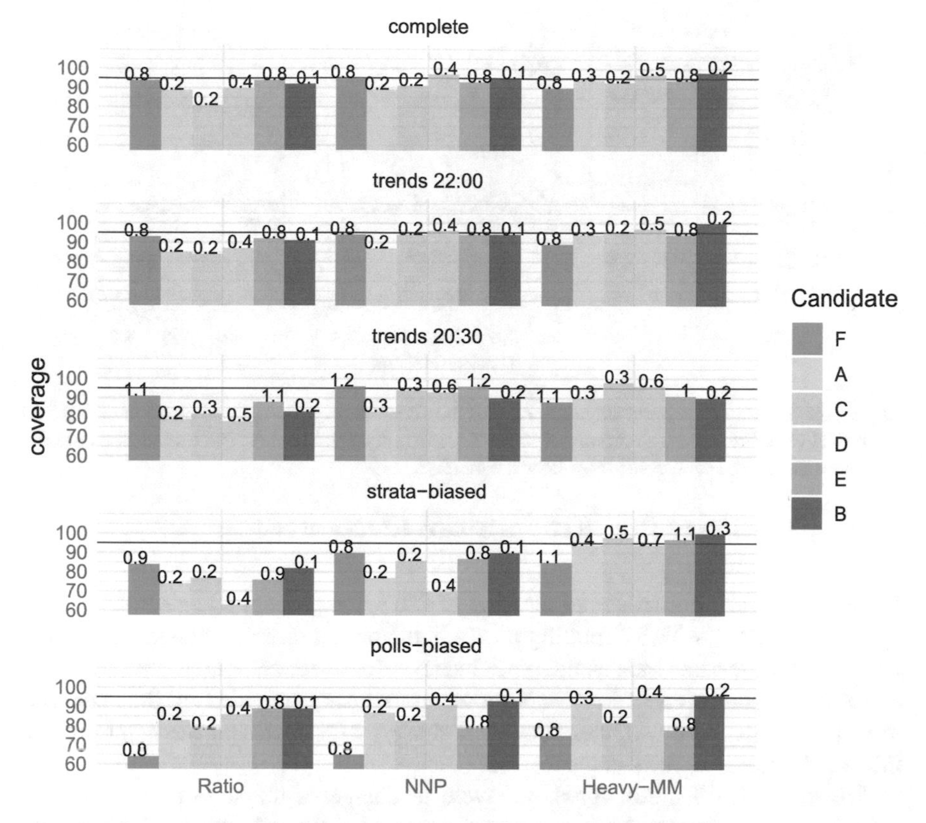

Fig. 4 Calibration of the ratio estimator, the NNP model, and the heavy-MM model. The size of the bars represents the percentage of times that the intervals cover the true value. The numbers above the bars indicate the mean of the 95% intervals precision

such as collapses or strata redistribution, to allow the methods to return an answer and provide more accurate estimations. Nevertheless, the heavy-MM model requires no special modification, as all cases are handled by standard Bayesian inference. Moreover, collapses or strata redistribution do not amend the bias. For this reason, the ratio estimator and the NNP model present a greater risk in the case where the partial sample used to present the final result has an important bias.

The refinement of the heavy-MM model was carried out with 2012 data from Guanajuato state and involved likelihood improvement to gross inadequacies. However, the final calibration was also corroborated with 2012 and 2006 data from Guanajuato, Chiapas, and Morelos. The most appealing part of the model is that, in the absence of a response, it attracts the parameters of a group toward the group mean. As a result, we have a consistent treatment of missing data in samples, and better interval coverage properties when the sample data is biased. However, a disadvantage is the heavy-MM model estimation method is much slower than the others. Running in parallel a candidate in each core, the heavy-MM model estimates with the complete sample in approximately 5 min, while the other two methods run in less than 30 s.

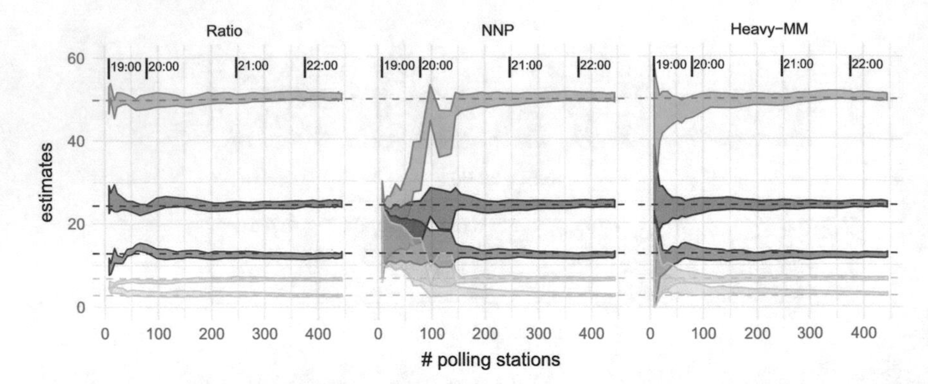

Fig. 5 Estimation of the percentage of votes of the Guanajuato gubernatorial election of 2018 as the partial samples arrived (intervals of 95% of probability/confidence). The dotted lines are the true percentages and each color represents a candidate

5 Application to the 2018 Mexican Elections

In the Guanajuato gubernatorial election of 2018, the probability intervals of 9:45 pm were reported, with 357 polling places in the partial sample. Figure 5 shows the monitoring of the intervals as the partial samples arrived, computed with the three methods we compared in Sect. 4. Note the consistency provided by the proposed model, the order of the winners does not change with time, and the length of the interval decreases as the sample arrives.

Within the 2018 quick count, we were in charge of the state of Guanajuato. Additionally, we were part of the support team for the state of Chiapas. To illustrate the generality of the model, Fig. 6 shows the monitoring of the intervals as the partial samples arrived for Chiapas. In fact, Chiapas is an interesting example as there was a close competition for the second and third places.

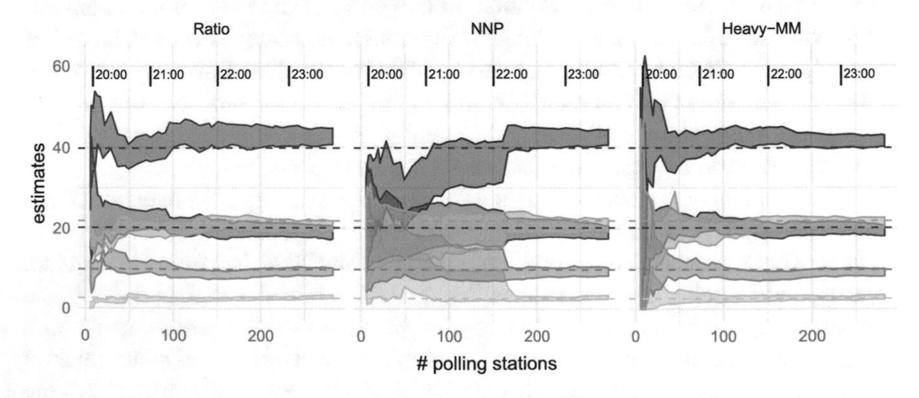

Fig. 6 Estimation of the percentage of votes of the Chiapas gubernatorial election of 2018 as the partial samples arrived (intervals of 95% of probability/confidence). The dotted lines are the true percentages and each color represents a candidate

6 Conclusions and Future Work

We presented the heavy-MM model as an alternative model to estimate the total number of votes in the Mexican elections. In this context, we have an incomplete and biased sample. The heavy-MM model provides: (1) a consistent treatment of missing data, (2) more stable behavior of partial samples, and (3) better coverage for certain types of bias in the samples.

Future work includes, in one direction, attempting a larger set of predictors with the 2018 data. We focused our attention on the predictors available in the INE databases. However, other information could be useful. Also, a deep analysis of the interactions could improve the predictions of the model. In another direction, it is worth trying to make the estimation method more efficient. For instance, modeling the covariance among the candidates could improve the coverage; however, it implies a much greater computational effort that requires a faster algorithm to maintain the needed threshold of 10 min.

References

1. COTECORA: Criterios científicos, logísticos y operativos para la realización de los conteos rápidos y protocolo de selección de las muestras (2018). https://repositoriodocumental.ine.mx/xmlui/bitstream/handle/123456789/96237/CGor201805-28-ap-27-a1.pdf
2. Gabry, J., Simpson, D., Vehtari, A., Betancourt, M., Gelman, A.: Visualization in Bayesian workflow (2017). arXiv preprint arXiv:1709.01449v5
3. Gelman, A., Hill, J.: Data Analysis Using Regression and Multilevel/Hierarchical Models. Cambridge University Press, Cambridge (2006)
4. Gelman, A., Stern, H.S., Carlin, J.B., Dunson, D.B., Vehtari, A., Rubin, D.B.: Bayesian Data Analysis. Chapman and Hall/CRC, United Kingdom (2013)
5. Ghosh, S.K., Mukhopadhyay, P., Lu, J.C.: Bayesian analysis of zero-inflated regression models. J. Stat. Plan. Inference **136**(4), 1360–1375 (2006). https://doi.org/10.1016/j.jspi.2004.10.008, http://www.sciencedirect.com/science/article/pii/S0378375804004008
6. Little, R.J.: Calibrated Bayes, an alternative inferential paradigm for official statistics. J. Off. Stat. **28**(3), 309 (2012)
7. Mendoza, M., Nieto-Barajas, L.E.: Quick counts in the Mexican presidential elections: a Bayesian approach. Electoral Stud. **43**, 124–132 (2016)
8. Ortiz, T.: quickcountmx: an R package to estimate election results from Mexico. https://github.com/tereom/quickcountmx (2018)
9. Park, D.K., Gelman, A., Bafumi, J.: Bayesian multilevel estimation with poststratification: state-level estimates from national polls. Polit. Anal. **12**(4), 375–385 (2004)
10. Pool, I.d.S., Abelson, R.P., Popkin, S.: Candidates, Issues, and Strategies. Cambridge, MA (1965)

Bayesian Estimation for the Markov-Modulated Diffusion Risk Model

F. Baltazar-Larios and Luz Judith R. Esparza

Abstract We consider the Markov-modulated diffusion risk model in which the claim inter-arrivals, claim sizes, premiums, and volatility diffusion process are influenced by an underlying Markov jump process. We propose a method for obtaining the maximum likelihood estimators of its parameters using a Markov chain Monte Carlo algorithm. We present simulation studies to estimate the ruin probability in finite time using the estimators obtained with the method proposed in this paper.

Keywords Ruin probability · Bayesian estimation · Markov-modulated

1 Introduction

Lu and Li [9] considered a two-state Markov-modulated risk model giving explicit formulas for non-ruin probabilities when the initial reserve was zero and when both claim sizes have exponential, Erlang, and a mixture of exponential distributions. As a generalization, Ng and Yang [10] considered the same model except the claim size distributions were phase-type.

For its part, Bäuerle and Kötter [3] considered the Markov-modulated diffusion risk reserve process. They used diffusion approximation to show its relation with classical Markov-modulated risk reserve processes. They also showed that increasing the volatility of the diffusion increases the probability of ruin.

Asmussen [1] provided the stationary distribution of a queueing system in terms of ruin probabilities for an associated process and solved the corresponding ruin problem. He also put special attention to Markov-modulated models.

F. Baltazar-Larios (✉)
Facultad de Ciencias, Universidad Nacional Autónoma de México, A.P. 20-726, 01000 CDMX
Mexico City, Mexico
e-mail: fernandobaltazar@ciencias.unam.mx

L. J. R. Esparza
Catedra CONACyT, Universidad Autonoma Chapingo, Texcoco, Mexico
e-mail: judithr19@gmail.com

I. Antoniano-Villalobos et al. (eds.), *Selected Contributions on Statistics and Data Science in Latin America*, Springer Proceedings in Mathematics & Statistics 301,
https://doi.org/10.1007/978-3-030-31551-1_2

Respect to the estimation, Guillou et al. [6] showed the strong consistency of the maximum likelihood estimator (MLE) for a Markov-modulated Poisson process (MMPP) driven loss process in insurance with several lines of business and fitted their model to real sets of insurance data using an adaptation of the EM algorithm. The same authors in [7] studied the statistical analysis of an MMPP, proving the strong consistency and asymptotic normality of a maximum split-time likelihood estimator and presented an algorithm to compute it.

In this paper, we will consider a Bayesian estimation of the parameters of a generalized risk model, the Markov-modulated diffusion risk model (MMDRM). We consider that the MMDRM has been observed only via their inter-arrivals claims and claim sizes in an interval time, so the data can be viewed as incomplete observations from a model with a tractable likelihood function. The full data set is a continuous record of the MMDRM and the underlying process. Then, we can find the maximum likelihood estimates using techniques of hierarchical grouping for classification of claim sizes and by applying a Gibbs sampler (GS) algorithm in the case of incomplete data. The main contribution of this work is to provide a Bayesian algorithm for estimating the parameters of the MMDRM in this scenario.

This paper is organized as follows. In Sect. 2, we present the MMDRM. The likelihood function is presented in Sect. 3, including the GS algorithm for the case of incomplete data. In Sect. 4, we present a simulation study for the method proposed, and we estimate the ruin probability in finite time. Final remarks are given in Sect. 5.

2 The Model

Let $\boldsymbol{U} = \{U_t\}_{t\geq 0}$ be the MMDRM where the underlying random environment is denoted by $\boldsymbol{J} = \{J_t\}_{t\geq 0}$, which is a homogeneous and irreducible Markov jump process (MJP), with finite state-space $E = \{1, \dots, m\}$, intensity matrix $\boldsymbol{Q} = \{q_{ij}\}_{ij=1}^{m}$ where $q_i = -q_{ii} = -\sum_{i\neq j} q_{ij}$, initial distribution $\boldsymbol{\pi}^0 = (\pi_1^0, \pi_2^0, \dots, \pi_m^0)$, and stationary distribution $\boldsymbol{\pi} = \{\pi_1, \dots, \pi_m\}$. We interpret each state of the MJP, $\boldsymbol{J}$, as a state of the economy. The MJP influences the premium rate, the arrival intensity of claims, the claim size distribution, and the volatility of the diffusion.

Let N_t be the number of claims in the time interval $[0, t]$ for all $t > 0$. We suppose that if $J_s = i$ for all $s \in [t, t+h]$ for some $i \in E$, $t \geq 0$, $h > 0$, then $N_{t+h} - N_t \sim \text{Poisson}(\lambda_i h)$, for $\lambda_i > 0$. Let $\boldsymbol{\lambda} = \{\lambda_1, \dots, \lambda_m\}$. We assume that the process $\boldsymbol{N} = \{N_t\}_{t\geq 0}$, called modulated Poisson process, has independent increments given the process $\boldsymbol{J}$.

On the other hand, let T_n, $n \in \mathbb{N}$, be the arrival time of n-th claim. Then, the amount of the n-th claim, X_n, given $J_{T_n} = i$, has a distribution function given by $F_i(\cdot|\eta_i)$, where F_i is assumed to belong to a family of distributions parametrized by η_i, and also F_i has support $(0, \infty)$, with density function f_i and mean $\mu_i < \infty$, for $i \in E$. Let $\boldsymbol{\mu} = \{\mu_1, \dots, \mu_m\}$, and $\boldsymbol{\eta} = \{\eta_1, \dots, \eta_m\}$. Let f_i be the corresponding density function of F_i, $i \in E$.

We will assume the claims sizes $\{X_n\}_{n\geq 1}$ are conditionally independent given $\boldsymbol{J}$. If $J_s = i$ for all $s \in [t, t+h]$ for some $i \in E$, then the volatility of the diffusion on the interval $[t, t+h]$ is given by $\sigma_i > 0$. Let $\boldsymbol{\sigma} = \{\sigma_1^2, \ldots, \sigma_m^2\}$.

We assume the premium income rate at time s is c_{J_s}, i.e., as long as $J_s = i$ we have a linear income stream at rate c_i. Then the surplus process $\{U_t\}_{t\geq 0}$ MMDRM is given by

$$U_t = u + \int_0^t c_{J_s} ds - \sum_{k=1}^{N_t} X_k + \int_0^t \sigma_{J_s} dW_s, \tag{1}$$

where $u \geq 0$ is the initial capital, and $\boldsymbol{W} = \{W_t\}_{t\geq 0}$ is a standard Brownian motion. Let $\boldsymbol{\theta} = (\boldsymbol{Q}, \boldsymbol{\lambda}, \boldsymbol{\eta}, \boldsymbol{\sigma})$ be the parameters of the MMDRM. The structure of the MMDRM differs from the classical risk model where trajectories are linear with jumps. However, the MMDRM arises as a limit of properly scaled classical risk processes, this means the MMDRM can be interpreted as a risk process with very small and frequent claims.

We define the ultimate ruin time as follows:

$$\tau = \inf\{t > 0 | U_t < 0\} \quad \text{if} \quad \{t > 0 | U_t < 0\} \neq \emptyset,$$

and $\tau = \infty$ if ruin never occurs.

The ruin probability is given by $\psi(u) = \mathbb{P}(\tau < \infty | U_0 = u)$, while the ruin probability in $[0, T]$ (finite time) is given by $\psi(u, T) = \mathbb{P}(\tau < T | U_0 = u)$, for $T > 0$. We also define the ultimate ruin probability given the initial state of the modulating process as follows:

$$\psi_i(u) = \mathbb{P}(\tau < \infty | U_0 = u, J_0 = i), \tag{2}$$

for all $i \in E$. By the law of total probability, we get

$$\psi(u) = \sum_{i\in E} \pi_i^0 \psi_i(u). \tag{3}$$

When studying the ultimate ruin probabilities, since we can apply the time change $\hat{X}_t := X_{S(t)}$ with $S(t) = \int_0^t \frac{1}{c_{J_s}} ds$, so without loss of generality, we can suppose that $c_{\{\cdot\}} = 1$ (see [3]).

In this paper, we will use the survival probability, and the survival probabilities given the initial state of $\boldsymbol{J}$, which are defined as $\phi(u) = 1 - \psi(u)$, and $\phi_i(u) = 1 - \psi_i(u)$, $i \in E$, respectively.

If $\kappa = 1 - \sum_{i\in E} \pi_i \lambda_i \mu_i \leq 0$, then for all $u \geq 0$ it holds that $\psi(u) = 1$ (see [3]). We will assume that $\kappa > 0$.

3 Estimation of the Parameters of the Markov-Modulated Diffusion Risk Model

In this section, we present a Bayesian algorithm for the estimation of $\boldsymbol{\theta}$. We consider the case where the only data available corresponds to the arrival times and claim sizes of the MMDRM in a time interval.

3.1 *Likelihood Function in the Full Data Case*

Let suppose the MMDRM has been observed continuously in the interval time $[0, T]$, $T > 0$. If we define $\boldsymbol{U}^c = \{U_t\}_{t\in[0,T]}$ and $\boldsymbol{J}^c = \{J_t\}_{t\in[0,T]}$, then the complete likelihood function can be written as follows:

$$L_T^c(\boldsymbol{\theta}) = p(\boldsymbol{U}^c, \boldsymbol{J}^c; \boldsymbol{\theta}) = p(\boldsymbol{U}^c|\boldsymbol{J}^c; \boldsymbol{\theta})p(\boldsymbol{J}^c; \boldsymbol{\theta})$$

$$= \prod_{i=1}^{m}\left[\left(\prod_{j=1}^{c_i}\lambda_i e^{-\lambda_i l_{ij}}\right)\left(\prod_{j=1}^{c_i} f_i(x_{ij}|\eta_i)\right)\left(\prod_{j=1}^{N_i}\prod_{k=2}^{r_{ij}}\frac{e^{-\frac{z_{ijk}^2}{2\sigma_i^2\Delta}}}{\sqrt{2\pi\sigma_i^2\Delta}}\right)\left(\prod_{j\neq i}^{m} q_{ij}^{N_{ij}} e^{-q_{ij}R_i}\right)\right], \quad (4)$$

where N_{ij} is the number of jumps from state i to j; N_i is the number of visits to state i; R_i is the total time spent in state i; l_{ij} is the inter-arrival time of the j-th claim while the process $\boldsymbol{J}$ stay in state i; x_{ij} is the amount of the j-th claim when $\boldsymbol{J}$ is in the state i; c_i is the number of claims when $\boldsymbol{J}$ is in the state i; $z_{ijk} = w_{ijk} - w_{ijk-1}$, where w_{ijk} is the k-th observation of $\boldsymbol{W}$ at the j-th visit of the process $\boldsymbol{J}$ at state i; and r_{ij} is the number of observations of the $\boldsymbol{W}$ at the j-th visit of the process $\boldsymbol{J}$ at state i. Since we cannot simulate continuous trajectories of a Brownian motion, we will simulate observations in each time interval of size $\Delta > 0$.

The maximum likelihood estimators of the parameters are given by

$$\hat{\lambda}_i = \frac{c_i}{\sum_{j=1}^{c_i} l_{ij}} \quad \hat{\sigma}_i^2 = \frac{\sum_{j=1}^{N_i}\sum_{k=2}^{r_{ij}} z_{ijk}^2}{\Delta\sum_{j=1}^{N_i}(r_{ij}-1)} \quad \hat{q}_{ij} = \frac{N_{ij}}{R_i}, \quad (5)$$

and the maximum likelihood estimator of $\boldsymbol{\eta}$ depends only on their corresponding density function.

3.2 *Discrete Observation of the Modulated Process*

Now, we consider that the MMDRM has been observed only via their claims inter-arrivals and claim sizes in the interval time $[0, T]$. Given $N_T = n$, we are interested in the inference of the parameter $\boldsymbol{\theta}$ based on $\{T_1, T_2, \ldots, T_n\}$, with $T_0 := 0$,

$\boldsymbol{U}^d = \{U_{T_0}, U_{T_1}, \ldots, U_{T_n}\}$, and $\boldsymbol{X} = \{x_1, x_2, \ldots, x_n\}$. Moreover, we suppose that m, u, J_{T_0} are known for all $i \in E$, and each f_i belongs to some parametric family, with their parameters η_i unknown.

To find the maximum likelihood estimators, we propose the following Bayesian algorithm.

Algorithm 1. General algorithm

1: Classify $\boldsymbol{X}$ into m groups and label each group according to its average, i.e., $x_j \in G_i$ if the average of G_i is $\hat{\mu}_i$, for $j = 1, \ldots, n, i \in E$, and $\hat{\mu}_1 < \hat{\mu}_2 < \cdots < \hat{\mu}_m$.
2: Estimate $\boldsymbol{\eta}_i$ based on G_i.
3: If $x_j \in G_i$ make $J^d_{T_j} = i$ for $j = 1, 2 \ldots, n$,
4: Estimate $\boldsymbol{Q}$ given $\boldsymbol{J}^d = \{J^d_{T_1}, \ldots, J^d_{T_n}\}$.
5: Given $\hat{\boldsymbol{Q}}$ we simulate a continuous path $J^c(T)$ of $\boldsymbol{J}$ in $[0, T]$. In particular, $J^c_{T_i}(T) = J^d_{T_j}$ for $j = 0, 1, \ldots, n$.
6: Estimate $\boldsymbol{\lambda}$ given $J^c(T)$.
7: Estimate $\boldsymbol{\sigma}$ given $J^c(T)$

After a burn-in period, we can obtain the $\boldsymbol{\theta}$ samples.

To implement this algorithm, in the following we will give a detailed description of each step. Step 3 is trivial.

Step 1

For this step, we can use techniques of hierarchical grouping, which start from as many clusters as there are elements in the database, and from that point, they integrate groups generally by a criterion of proximity until the grouping constitutes a single conglomerate.

Some well-known procedures of the hierarchical grouping are the linking intergroups, intragroups, the nearest neighbor, thc furthest neighbor, grouping of centroids, grouping of medians, and the Ward. In this work, we will use the latter. Ward [13] states that the procedure allows forming mutually exclusive hierarchical groups based on similarity with respect to specification characteristics.

We consider the traditional mixture modeling approach that requires the number of clusters to be specified in advance, i.e., m. We recommend the reader to check [8] for more classification and clustering methods viewed from statistical learning.

Step 2

Based on the classification of Step 1, in this step, we obtain the maximum likelihood estimators for η_i using $\boldsymbol{x}_i = \{x_{1,i} \ldots, x_{c_i,i}\}$ and the correspondent density function f_i for $i = 1, 2, \ldots, m$.

Step 4

In Step 3, the underlying MJP, $\boldsymbol{J}$, is recorded at discrete times. We can think of $\{J^d_{T_j}\}$ as incomplete observations of the full data set given by the sample path $\{J_t\}_{t\in[0,T]}$. We are interested in the inference of the intensity matrix $\boldsymbol{Q}$ based on the sample observation of $\boldsymbol{J}^d$.

The likelihood function for the discrete-time data $\boldsymbol{J}^d$ is given in terms of transition probabilities:

$$L^d_T(\boldsymbol{Q}) = \prod_{j=2}^{n} p_{J_{T_{j-1}} J_{T_j}}(S_j), \tag{6}$$

where $S_j = T_j - T_{j-1}$. The difficulty lies in finding the derivative of (6) with respect to its entries, which has such a complicated form that the null cannot be found analytically. Hence, no analytical expression for the maximum likelihood estimator with respect to $\boldsymbol{Q}$ is available. Moreover, there are some other issues that were discussed in [4]: it is possible that the maximum does not exist, problems of identifiability and of existence, and uniqueness of the estimator, among others.

In [4], the authors provided a likelihood inference for discretely observed Markov jump processes with finite state, the existence and uniqueness of the maximum likelihood estimator of the intensity matrix, and demonstrated that the maximum likelihood estimator can be found either by the EM algorithm or by a Markov chain Monte Carlo (MCMC) procedure. Based on [4], the Gibbs sampler works as follows.

Algorithm (a) Gibbs sampler for estimating $\boldsymbol{Q}$

1: Draw an initial $\boldsymbol{Q}$ from the prior.
2: Simulate a Markov jump process J with intensity matrix $\boldsymbol{Q}$ up to time T such that $J_{T_j} = i$, for all $j = 1, \ldots, n$ and $i \in \{1, \ldots, m\}$.
3: Calculate the statistics N_{ij} and R_i from $\{J(t)\}_{0\le t\le T}$.
4: Draw a new $\boldsymbol{Q}$ from the posterior distribution.
5: Go to 2.

After a burn-in period, we can obtain the $\boldsymbol{Q}$ samples.

For Step 1 of this algorithm, the prior is given by

$$p(\boldsymbol{Q}) \propto \prod_{i=1}^{m} \prod_{j\neq i} q_{ij}^{a_{ij}-1} e^{-q_{ij} b_i}, \tag{7}$$

where $a_{ij} > 0$ and $b_i > 0$, $i, j \in E$, are constants to be chosen conveniently (see [4]). Then $q_{ij} \sim Gamma(a_{ij}, b_i)$.

For the simulation of conditioned paths of the MJP in Step 2 of Algorithm (a), we use the algorithm proposed in [2].

Finally, the posterior distribution is given by

$$p^*(\boldsymbol{Q}) = L_T^c(\boldsymbol{\theta})p(\boldsymbol{Q}) \propto \prod_{i=1}^{m}\prod_{j\neq i} q_{ij}^{N_{ij}+a_{ij}-1} e^{-q_{ij}(R_i+b_i)},$$

then, $q_{ij} \sim Gamma(N_{ij} + a_{ij}, R_i + b_i)$.

Step 5

When Algorithm (a) converges, we use the estimated $\hat{\boldsymbol{Q}}$ in order to generate a continuous path $J^c(T)$ such that $J^c_{T_j}(T) = J^d_{T_j}$ for $j = 0, 1, \ldots, n$.

Step 6

Let m_i be the number of claims arriving while the underlying process $J^c(T)$ is in state i, $i \in E$. We define $T_{k_i} := T_j$ if $J_{T_j} = i$, for $k_i = 1, 2, \ldots, m_i$ and $j = 1, 2 \ldots, n$. Also the number of jumps of $J^c(T)$ in $[T_{j-1}, T_j]$ is zero.

If

$$S_{k_i} := T_{k_i+1} - T_{k_i} \sim \text{Exp}(\lambda_i) \qquad \text{for} \qquad k_i = 1, 2, \ldots, m_i,$$

then, the estimator of λ_i, $i \in E$, is given by

$$\hat{\lambda}_i = \frac{m_i}{\sum_{k=1}^{m_i} S_{k_i}}.$$

Step 7

Now, to find an estimator of $\boldsymbol{\sigma}$, based on [11], we will use data augmentation for diffusion. Suppose that we observe the Brownian motion $\boldsymbol{W}^i$, $i = 1, \ldots, m$, with drift parameter zero and diffusion parameter $\sigma_i > 0$. In this case, we have only discrete-time observations of $\boldsymbol{W}^i$ at claim times. Then, we define

$$W^+_{k_i} := U_{T_{k_i}} + x_{T_{k_i}} - c_i S_{k_i} \qquad W^-_{k_i} := U_{T_{k_i-1}}, \tag{8}$$

where $T_{0_i} := T_{1_i} - S_{1_i}$, $W^+_{k_i}$ is the insurance capital without prime just before of the k-th claim without jumps in $J^c(T)$, and $W^-_{k_i}$ is the insurance capital at time $T_{k_i} - S_{k_i}$ for $k_i = 1, 2, \ldots, m_i$, and $i \in E$. Then, we can think of the data set $\{W^-_1, W^+_1, \ldots, W^-_{m_i}, W^+_{m_i}\}$ as incomplete observation of a full data set given the sample Brownian motion $\boldsymbol{W}^i_t$ with volatility σ_i, $t \in [T_{k_i-1}, T_{k_i}]$ for $k_i = 1, 2, \ldots, m_i$. Then, we propose the following Gibbs sampler algorithm to find the maximum likelihood estimator of $\boldsymbol{\sigma}$. We define $\zeta_i = \frac{1}{\sigma_i^2}$ for all $i \in E$.

Algorithm (b) Gibbs sampler algorithm for estimating σ_i^2.

1: Draw ζ_i from the prior distribution $Exp(\zeta_i^0)$ and make $\sigma_i = \frac{1}{\sqrt{\zeta_i}}, i \in E$.
2: Simulate sample paths of W_t^i for $t \in [T_{k_i-1}, T_{k_i}]$ with $i \in E$ and $k_i = 1, \ldots, m_i$. For this, we generate a Brownian bridge $(T_{k_i-1}, W_{k_i}^-, T_{k_i}, W_{k_i}^+) \approx \{w_{k_i0} = W_{k_i}^-, w_{k_i1}, \ldots, w_{k_i\Delta_{k_i}} = W_{k_i}^+\}$, where $\Delta_{k_i} := \lfloor S_{k_i}/\Delta \rfloor$, and Δ is the discretization of the Brownian motion.
3: Draw ζ_i from the posterior distribution

$$\text{Gamma}\left(1 + \frac{m_i(\Delta_{k_i} - 1)}{2}, \zeta_i^0 + \frac{\sum_{k=1}^{m_i}\sum_{\ell=1}^{\Delta_{k_i}}(w_{k_i\ell} - w_{k_i(\ell-1)})^2}{2\Delta\sum_{k=1}^{m_i}(\Delta_{k_i} - 1)}\right).$$

4: Go to 2.

After a burn-in period, we can obtain the $\boldsymbol{\sigma}$ samples.

4 A Simulation Study

In this section, we apply the method developed before to an MMDRM simulated at time interval [0, 1000], when the underlying MJP has the state-space $E = \{1, 2, 3\}$, initial probability vector $\boldsymbol{\pi}^0 = (1/2, 1/4, 1/4)$, and intensity matrix

$$\boldsymbol{Q} = \begin{pmatrix} -2 & 1 & 1 \\ 1 & -2 & 1 \\ 1 & 1 & -2 \end{pmatrix}.$$

The conditional distribution of claim inter-arrivals is $S_n | J_{T_n} = i \sim \text{Exp}(\lambda_j)$; with $\boldsymbol{\lambda} = (0.1, 0.14, 0.09)$, the conditional distribution of claim sizes is $X_n | J_{T_n} = i \sim Gamma(\alpha_i, \beta_i)$ with $\boldsymbol{\alpha} = (1, 15, 40)$, $\boldsymbol{\beta} = (3, 3, 3)$, and the diffusion parameters are $\boldsymbol{\sigma} = (0.9, 0.8, 0.5)$. We consider discrete observations of the surplus process at each inter-arrival point. We estimate the parameters and we also report their standard errors (SD). Moreover, we present a simulation study for the estimated probability of ruin using the real parameter $\boldsymbol{\theta}$ and its estimation $\hat{\boldsymbol{\theta}}$.

4.1 Estimation of Parameters

According to Algorithm 1, the steps are the following:

Step 1

First, classify the observed claims into three groups. The corresponding result is presented in Fig. 1.

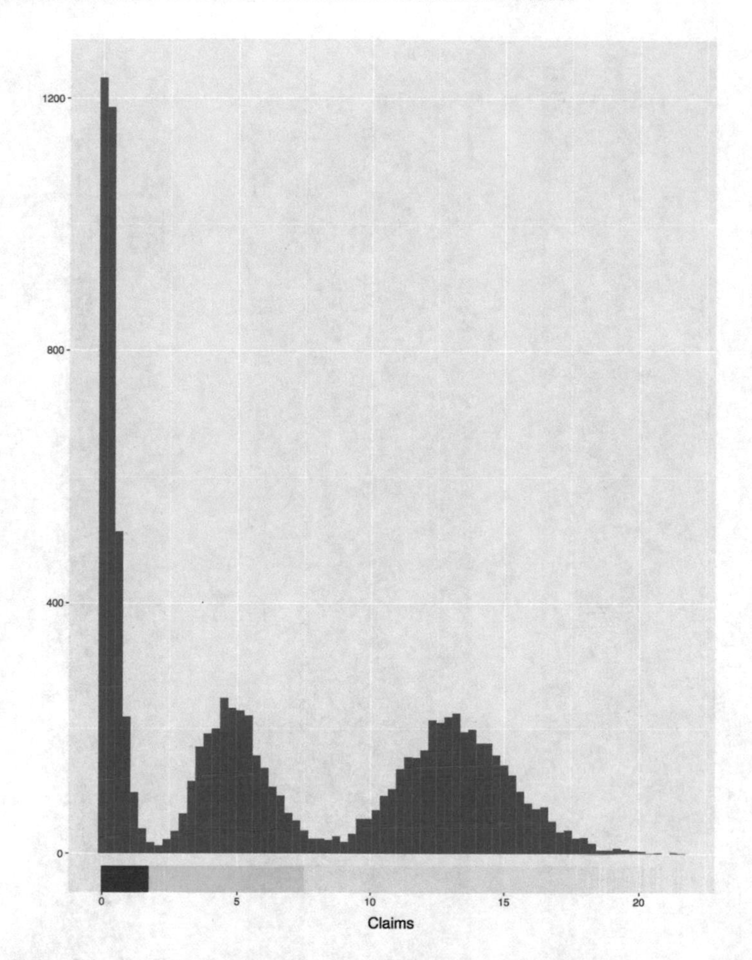

Fig. 1 Claims Distribution. Classification of the data into three groups

Table 1 Maximum likelihood estimators for $\boldsymbol{\alpha}$ and $\boldsymbol{\beta}$

Parameter	Real value	Estimation	SD	Parameter	Real value	Estimation	SD
α_1	1	1.02362	0.02213365	β_1	3	3.205170	0.08841398
α_2	15	14.60547	0.42068222	β_2	3	2.967807	0.08696558
α_3	40	39.010476	0.90507887	β_3	3	2.931284	0.06844663

Step 2

We estimated the parameters of $f_i, i = 1, 2, 3$, of the Gamma distribution according to the classification given before. The results are given in Table 1.

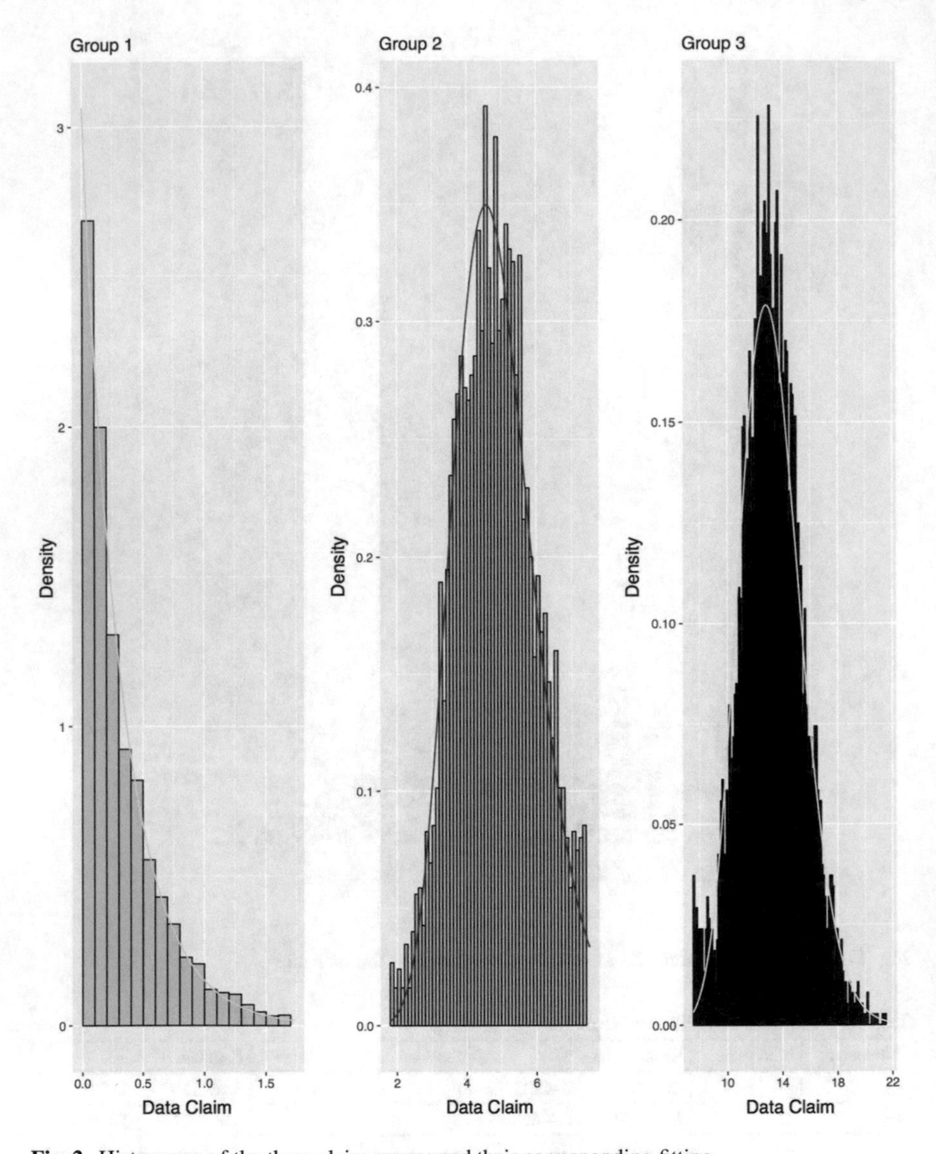

Fig. 2 Histograms of the three claim groups and their corresponding fitting

Figure 2 shows the histogram for the three sets of claims with distribution $Gamma(\alpha_i, \beta_i)$ for $i = 1, 2, 3$ and their corresponding fitting.

Step 4

We estimate the parameters of the intensity matrix $\boldsymbol{Q}$ considering the prior distribution (7) with $a_{ij} = 1$ and $b_i = 1$ for all $i, j \in E$ (see [4]).

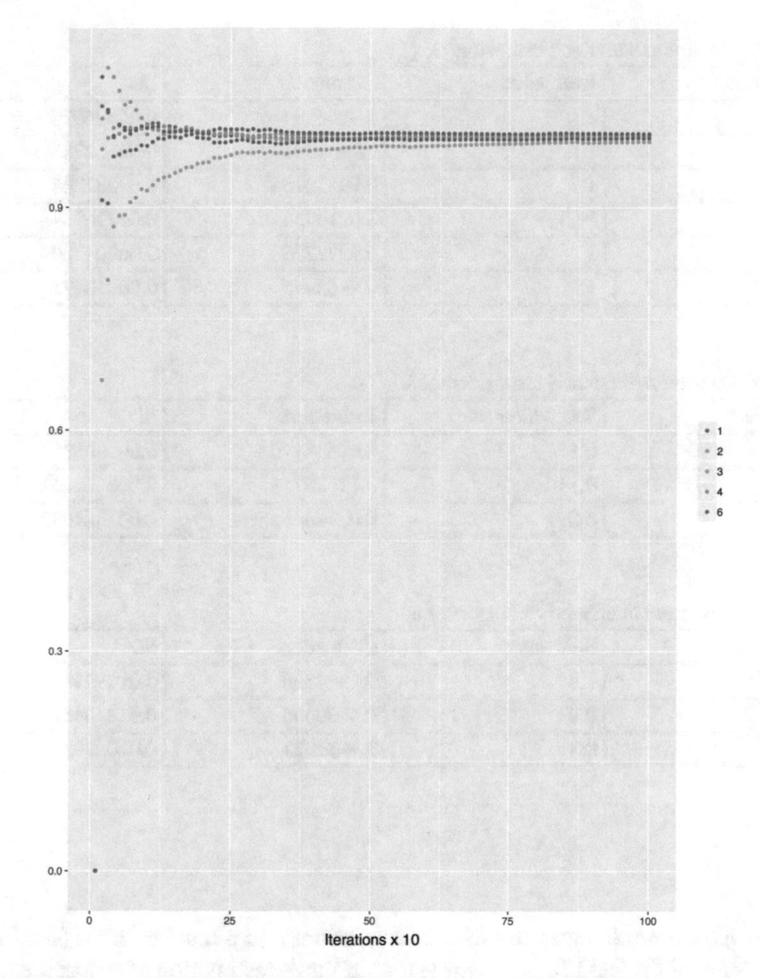

Fig. 3 Ergodic averages taken every 10 iterations, of each $q_{ij}, i, j \in E, i \neq j$

Algorithm (a) was run with 1000 iterations and a burn-in of 300 based on Fig. 3 where we plot the ergodic averages for each of the 10 iterations.

The average and standard deviation of the last 700 iterations are presented in Table 2.

Step 6

Having $\hat{\boldsymbol{Q}}$, in Table 3 we show the corresponding estimations of $\boldsymbol{\lambda}$.

Figure 4 shows the histograms of the times classified into the three groups corresponding to $\lambda_i, i = 1, 2, 3$.

Table 2 Maximum likelihood estimators of $\boldsymbol{Q}$

Parameter	Real value	Estimation	SD
q_{12}	1	0.9966584	0.05406979
q_{13}	1	0.9954521	0.06162783
q_{21}	1	0.9953315	0.07231265
q_{23}	1	1.0016352	0.05726284
q_{31}	1	0.9972229	0.06204729
q_{32}	1	0.9963995	0.06310475

Table 3 Maximum likelihood estimators for $\boldsymbol{\lambda}$

Parameter	Real value	Estimation	SD
λ_1	0.1	0.09638170	0.001365332
λ_2	0.14	0.13685713	0.002472661
λ_3	0.09	0.08748555	0.001281092

Table 4 Maximum likelihood estimators of $\boldsymbol{\sigma}$

Parameter	Real value	Estimation	SD
σ_1^2	0.9	0.8904599	0.01892117
σ_2^2	0.8	0.8090435	0.01334441
σ_3^2	0.5	0.4929025	0.01087935

Step 7

Finally, Algorithm (b) was run with 200 iterations and a burn-in of 50 (see Fig. 5). The average of the last 150 iterations was used for the estimation of $\boldsymbol{\sigma}$, and the results are given in Table 4.

4.2 Estimation of Ruin Probability for the MMDRM

Now, we estimate the ruin probability by a Monte Carlo (MC) method using the real and estimated parameter. Since our obtained estimators are based on observations in a finite time interval, in this section we study their precision when estimating the ruin probability in finite time using the true parameters and their estimators. The equivalent study for infinite time horizon would be by computing the exactly ultimate ruin probability (if possible) using the corresponding estimators and parameters. The MC method was run with a sample of size 1000, $\boldsymbol{\theta}$ and $\hat{\boldsymbol{\theta}}$. Figure 6 shows the ruin probability considering $\Delta = 0.1$, $T = 1$, and $u = 0, \ldots, 9$.

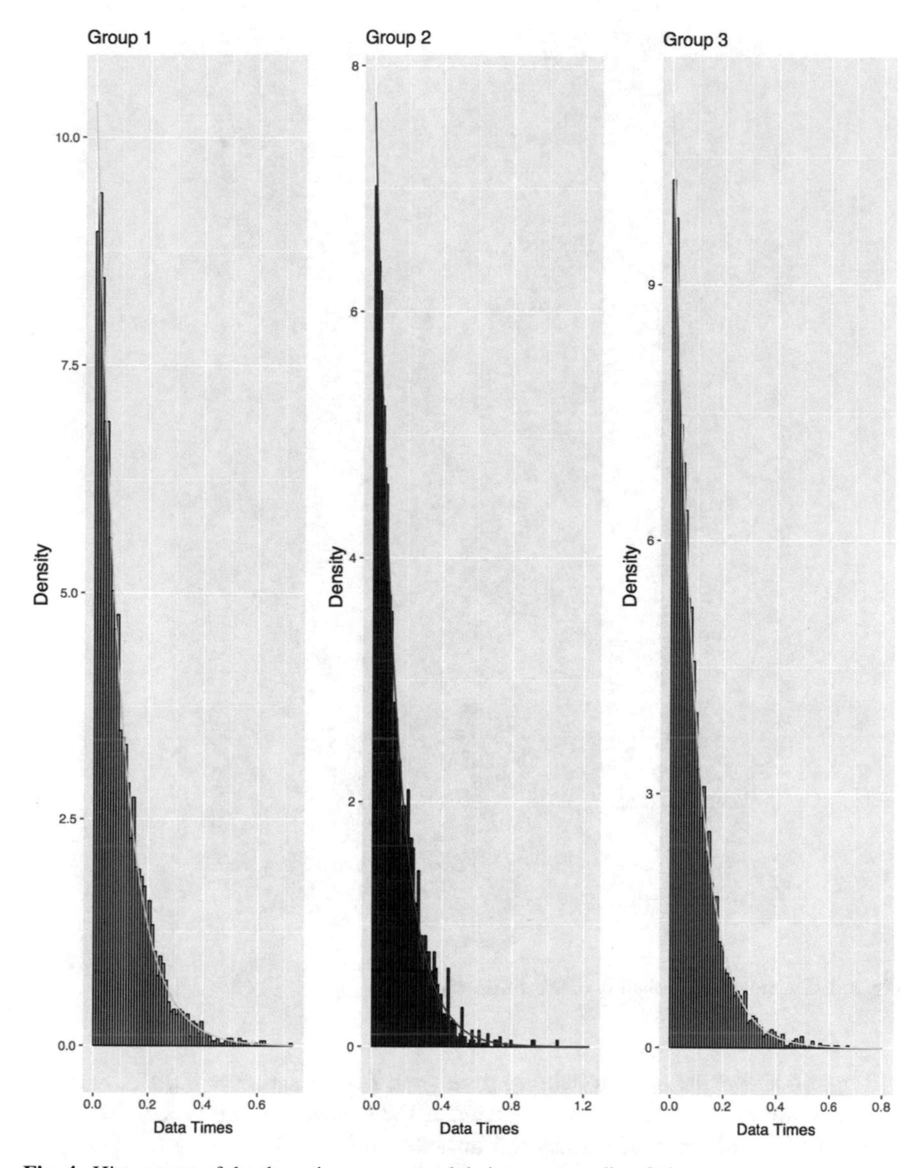

Fig. 4 Histograms of the three time groups and their corresponding fitting

Figure 7 shows the ruin probability considering $u = 0$, $\Delta = 0.1$, and $T = 1, 5, 10, 15, 20, 25, 30, 35, 40, 45$.

Another parameter to consider for a good simulation of the ruin is the length between the observations of the diffusion process, Δ. The best way to exemplify this is in the case of $u = 0$, where the instantaneous ruin is due to the Brownian motion becomes negative within a very small time interval.

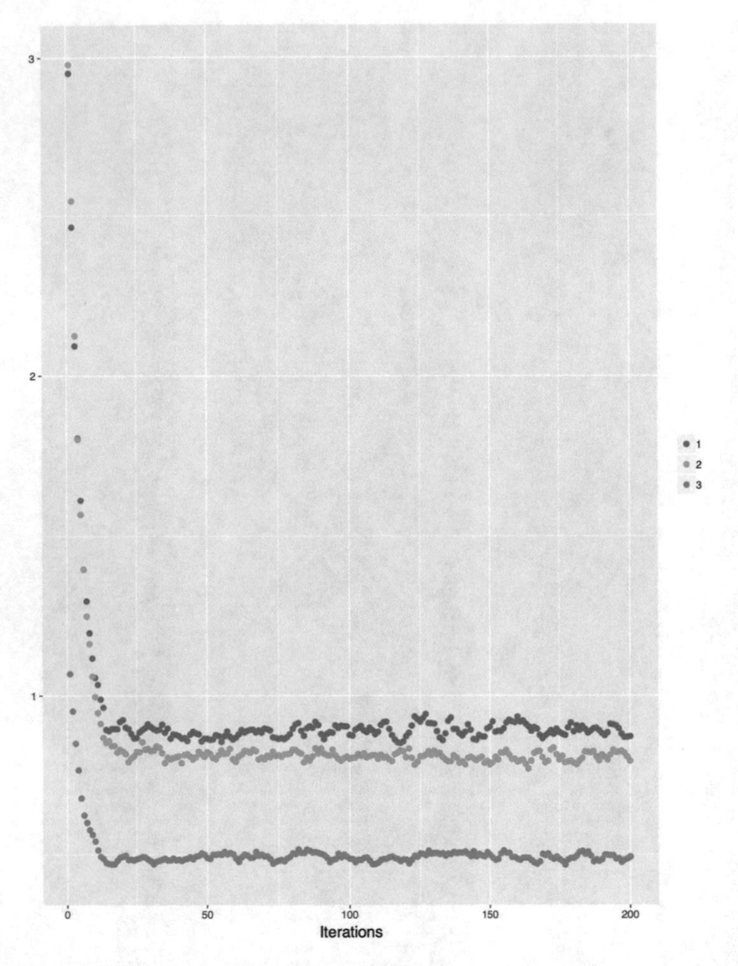

Fig. 5 Estimation of the volatility of the diffusions σ

Figure 8 shows the ruin probability for $u = 0$, $T = 1$, and $\Delta = 0.001, \ldots, 0.1$. Small values of Δ allow to observe better the behavior of the process with diffusion, and in this particular case, the instantaneous ruin by having initial capital zero.

5 Conclusions

We have presented a Gibbs sampler algorithm for obtaining maximum likelihood estimates of the parameters of the MMDRM when the underlying Markov jump process influences the premium rate, the arrival intensity of the claims, the claim size distribution, and the volatility of the diffusion. We considered the case when we have incomplete observations of the continuous-time records of the surplus and the

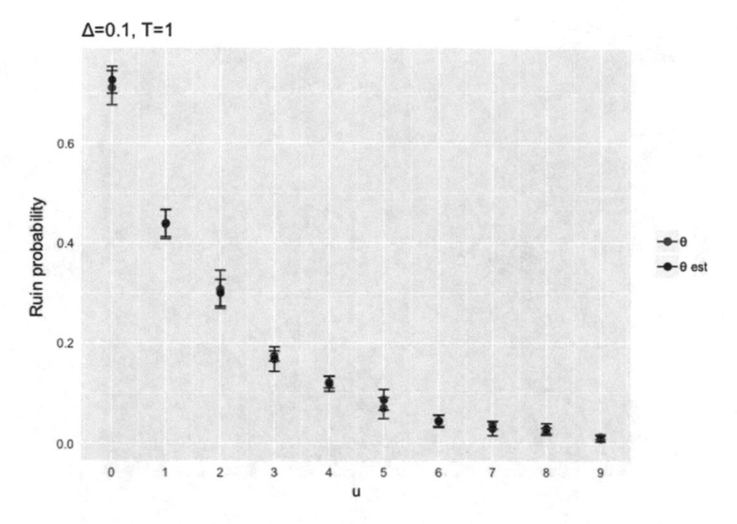

Fig. 6 Estimation of the ruin probability considering $\Delta = 0.1$, $T = 1$, and $u = 0, \ldots, 9$

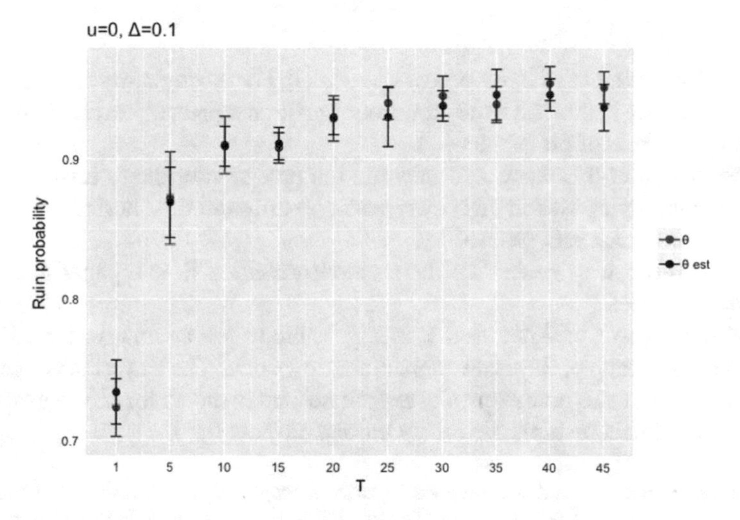

Fig. 7 Estimation of the ruin probability considering $u = 0$, $\Delta = 0.1$, and $T = 1, 5, 10, 15, 20, 25, 30, 35, 40, 45$

underlying Markov jump process. The data are discrete-time samples of the surplus process at the time each claim arrives.

We calibrated our algorithm with a simulation study with efficient results. Furthermore, we compared the estimation of the ruin probability in finite time using the real parameters and the estimators obtained by the proposed method in a Monte Carlo simulation. Using these estimators, it is possible to calculate or estimate the ruin probability in finite time. In our example, the claim size distribution belongs to the Phase-type class, so the ruin probabilities can be estimated by applying the

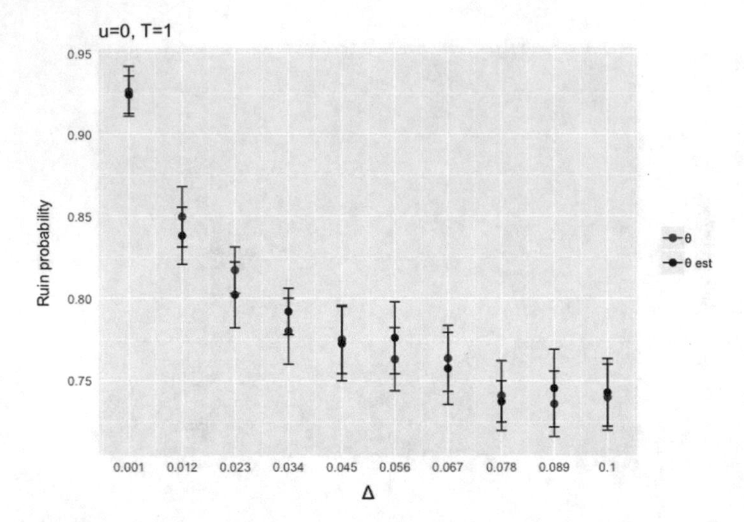

Fig. 8 Estimation of the ruin probability considering $u = 0$, $T = 1$, and $\Delta = 0.001, \ldots, 0.1$

Erlangization method to the risk process (see [12]). For those claim size distributions that do not belong to this class, the proposed algorithm represents an efficient method to estimate the ruin of probability.

An extension of this work is to measure the precision of the estimates obtained with the method proposed in this paper based on different time horizons in order to estimate the ultimate ruin probability.

It is not difficult to generalize the method presented in this paper using other diffusion processes.

Finally, in Step 1 of Algorithm 1, it is possible to use alternative techniques of clustering, for example, Bayesian nonparametric models. The Bayesian nonparametric approach estimates how many clusters are needed to model the observed data and allows future data to exhibit previously unseen clusters ([5]).

Acknowledgements Luz Judith Rodriguez Esparza is supported by a Catedra CONACyT. The research of F. Baltazar-Larios was supported by PAPIIT-IA105716. Both authors are thankful to the reviewers for their invaluable comments and suggestions, which improve the paper substantially.

References

1. Asmussen, S.: Stationary distributions via first passage times. In: Dshalalow, J.H. (ed.) Advances in queueing: Theory, methods, and open problems, pp. 79–102. CRC Press, Boca Raton (1995)
2. Asmussen, S., Hobolth, A.. Markov bridges, bisection and variance reduction. In: Plaskota, L., Wozniakowski, H. (eds.), Monte Carlo and Quasi-Monte Carlo Methods 2010, Springer Proceedings in Mathematics and Statistics, vol. 23, p. 322. Springer, Berlin (2012)

3. Bäuerle, N., Kötter, M.: Markov-modulated diffusion risk models. Scand. Actuar. J. **1**, 34–52 (2007)
4. Bladt, M., Sorensen, M.: Statistical inference for discretely observed markov jump processes. J. R. Stat. Soc. **67**(3), 395–410 (2005)
5. Gershmana, S.J., Blei, D.M.: A tutorial on Bayesian nonparametric models. J. Math. Psychol. **56**, 1–12 (2012)
6. Guillou, A., Loisel, S., Stupfler, G.: Estimation of the parameters of a Markov-modulated loss process in insurance. Insur. Math. Econom. **53**, 388–404 (2013)
7. Guillou, A., Loisel, S., Stupfler, G.: Estimating the parameters of a seasonal Markov-modulated Poisson process. Stat. Methodol. **26**, 103–123 (2015)
8. James, G., Witten, D., Hastie, T., Tibshirani, R.: An Introduction to Statistical Learning with Applications in R. Springer, New York (2014)
9. Lu, Y., Li, S.: On the probability of ruin in a Markov-modulated risk model. Insur.: Math. Econ. Elsevier. **37**(3), pp. 522–532 (2005)
10. Ng, A.C.Y., Yang, H.: On the joint distribution of surplus before and after ruin under a Markovian regime switching model. Stoch. Process. Their Appl. **116**(2), 244–266 (2006)
11. Roberts, G.O., Stramer, O.: On inference for partially observed nonlinear diffusion models using the metropolis-hastings algorithm. Biometrika Trust. **88**(3), 603–621 (2001)
12. Stanford, D.A., Yu, K., Ren, J.: Erlangian approximation to finite time ruin probabilities in perturbed risk models. Scand. Actuar. J. **1**, 38–58 (2011)
13. Ward, J.H.: Hierarchical grouping to optimize an objective function. J. Am. Stat. Assoc. **58**(301), 236–244 (1963)

Meta-Analysis in DTA with Hierarchical Models Bivariate and HSROC: Simulation Study

Sergio A. Bauz-Olvera, Johny J. Pambabay-Calero, Ana B. Nieto-Librero and Ma. Purificación Galindo-Villardón

Abstract In meta-analysis for diagnostic test accuracy (DTA), summary measures such as sensitivity, specificity, and odds ratio are used. However, these measures may not be adequate to integrate studies with low prevalence, which is why statistical modeling based on true positives and false positives is necessary. In this context, there are several statistical methods, the first of which is a bivariate random effects model, part of the assumption that the logit of sensitivity and specificity follow a bivariate normal distribution, the second, refers to the HSROC or hierarchical model, is similar to bivariate, with the particularity that it directly models the sensitivity and specificity relationship through cut points. Using simulations, we investigate the performance of hierarchical models, varying their parameters and hyperparameters and proposing a better management of variability within and between studies. The results of the simulated data are analyzed according to the criterion of adjustment of the models and estimates of their parameters.

Keywords Diagnostic precision · Meta-analysis · HSROC model · Bivariate model · Low prevalence

S. A. Bauz-Olvera (✉)
Facultad de Ciencias de la Vida, Escuela Superior Politécnica del Litoral, Guayaquil, Ecuador
e-mail: serabauz@espol.edu.ec

J. J. Pambabay-Calero
Facultad de Ciencias Naturales y Matemáticas, Escuela Superior Politécnica del Litoral, Guayaquil, Ecuador
e-mail: jpambaba@espol.edu.ec

A. B. Nieto-Librero · Ma. P. Galindo-Villardón
Dpto. de Estadística, Facultad de Medicina, Universidad de Salamanca, Salamanca, Spain
e-mail: ananieto@usal.es

Ma. P. Galindo-Villardón
e-mail: pgalindo@usal.es

Instituto de Investigación Biomédica (IBSAL), Salamanca, Spain

I. Antoniano-Villalobos et al. (eds.), *Selected Contributions on Statistics and Data Science in Latin America*, Springer Proceedings in Mathematics & Statistics 301,
https://doi.org/10.1007/978-3-030-31551-1_3

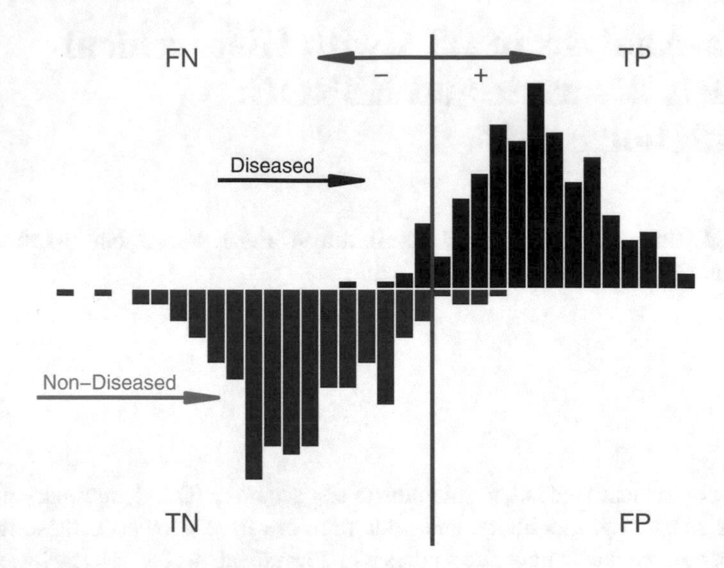

Fig. 1 Test threshold and impact on diagnostic accuracy

1 Introduction

One way to summarize the behavior of a diagnostic test from multiple studies is by calculating the mean sensitivity and the mean specificity, but these summary measures are not valid if there is heterogeneity between the studies and unfortunately the tests to detect heterogeneity are not very powerful [1], so that hierarchical models are necessary, which are capable of capturing this heterogeneity. In meta-analysis, the objective is to integrate the results of the different articles in which a diagnostic test is evaluated.

The starting information is a tetrachoric table in which measures such as true positive (TP), false positive (FP), false negative (FN), and true negative (TN) appear. Table 1 shows the concordance between the result of the test in binary form and the information on the disease (Fig. 1).

While the statistical methods for the meta-analysis of clinical trials are well developed and understood today, there are still challenges when performing meta-analysis of data from studies on the accuracy of the diagnostic test. This is mainly due to the bivariate nature of the response, where information on sensitivity and specificity should be summarized.

Table 1 Generic representation of the precision analysis of a diagnostic test

Patient's condition		
Result of the test	Positive	Negative
Positive	TP	FP
Negative	FN	TN

This correlation can be induced by populations or heterogeneous environments in the different trials, but it is mainly derived from the balance between sensitivity and specificity when the threshold to define the positivity of the test varies.

1.1 Hierarchical Models

More rigorous approaches have been proposed statistically based on hierarchical models that overcome the limitations of the method proposed by [2]. This section briefly describes the bivariate model [3] and the hierarchical model (HSROC) [4]. Both hierarchical models imply statistical distributions in two levels. At the lower level, the count of the values taken by the tetrachoric tables extracted from each study is modeled using binomial distributions and logistic transformations. At the higher level, it is assumed that the random effects of the study explain the heterogeneity between the studies beyond what is explained by the sampling variability at the lower level. The bivariate model and the HSROC model are mathematically equivalent when covariates are not available [5, 6] but differ in their parameterizations.

The Bivariate parameterization models the sensitivity, the specificity and the correlation between them directly. While the parameterization of the model HSROC models functions of thresholds of positivity, precision, and shape of the curve to define an SROC graph.

1.2 Bivariate Model

The bivariate model uses a different starting point for the ordered pairs of sensitivity and specificity of the meta-analyzes. Instead of transforming these two measurements into a diagnostic indicator of accuracy, as in the SROC approach, the bivariate model preserves the two dimensions of the data through analysis. It is assumed that the sensitivity values of the individual studies (after the logit transformation) of a meta-analysis follow a normal distribution around the value of the mean and with a certain amount of variability around the same mean. This is a random effects approach, similar to that used in therapeutic trials, to incorporate the variability not explained in the analysis. This variation in the underlying sensitivity between the studies may be related to the differences that persist in the study population, the implicit differences in the threshold, or imperceptible variations in the clinical trial protocol.

The same considerations apply to the specificities of these studies. The potential presence of a (negative) correlation between sensitivity and specificity within the studies is addressed by the explicit incorporation of this correlation in the analysis. The combination of normal distribution and logit transformations of sensitivity and specificity, recognize the possible correlation between them, leading to a normal bivariate distribution [7, 8].

The Bivariate method models sensitivity and specificity directly. It can be considered that the model has two levels corresponding to the variation within and between the studies. In the first level, it is assumed that the variability within the study for sensitivity and specificity follows a binomial distribution [9] have shown that a binomial probability distribution should be used to model variability within studies (especially when data are scarce). A more formal definition is the following.

A and B are denoted for sensitivity and specificity, respectively, and the number of true positives for y_{Ai} in study i, where n_{Ai} denote the total number of sick individuals and π_{Ai} represents the probability that the test will give a positive result in study i. Similarly, y_{Bi} the number of true negatives, where n_{Bi} denotes the total of healthy individuals and π_{Bi} is the probability that the test will give a negative result in study i, that is:

$$y_{Ai} \sim Binomial(n_{Ai}, \pi_{Ai}), y_{Bi} \sim Binomial(n_{Bi}, \pi_{Bi}) \quad (1)$$

The specificity pair, sensitivity for each study should be modeled jointly at level one (lower) since these measures are linked by study characteristics including the threshold of positivity. At the upper level, it is assumed that the logit transformations of the sensitivities have a normal with mean μ_A and variance σ_A^2, while the logit transformations of the specificities are distributed by a normal with mean μ_B and σ_B^2. Its correlation is included by modeling at the same time a single bivariate normal distribution, which allows the joint analysis of the sensitivity and specificity, which is a linear mixed model:

$$\begin{pmatrix} \mu_{Ai} \\ \mu_{Bi} \end{pmatrix} \sim N\left(\begin{pmatrix} \mu_A \\ \mu_B \end{pmatrix}, \Sigma\right), \Sigma = \begin{pmatrix} \sigma_A^2 & \sigma_{AB} \\ \sigma_{AB} & \sigma_B^2 \end{pmatrix} \quad (2)$$

where σ_A^2 and σ_B^2 denote the variability between the logit transformations of sensitivity and specificity, respectively, and σ_{AB} denotes the covariance between the logit transformations of sensitivity and specificity. The model can also be parameterized using the correlation $\rho_{AB} = \sigma_{AB}/(\sigma_A\ \sigma_B)$, which may be more interpretable than covariance. Therefore, the bivariate model without covariations has the following five parameters: μ_A, μ_B, σ_A^2, σ_B^2, and σ_{AB} (o ρ_{AB}). The Bivariate statistical model can be simplified assuming covariances or correlations equal to zero (that is, an independent variance–covariance structure), then the model is reduced to two univariate random effects logistic regression models for sensitivity and specificity.

1.3 Model HSROC

The HSROC model represents a general framework for the meta-analysis of trial accuracy studies and can be seen as an extension of the SROC approach of [2] in which rates of true positives (TTP) and false positive rate (TFP) for each study are directly modeled [10]. The HSROC model is a generalized mixed nonlinear model and takes the following form:

$$logit(\pi_{ij}) = (\theta_i + \alpha_i dis_{ij})exp(-\beta dis_{ij}) \quad (3)$$

where π_{ij} is the probability that a patient in study i with a disease condition j will obtain a positive test result. The disease state is represented by dis_{ij}, which is coded with -0.5 ($j = 0$) for individuals who do not possess the disorder (disease) and 0.5 ($j = 1$) for the group of individuals with the disease for the i-th study.

The implicit threshold θ_i (threshold parameter or positivity criterion) and α_i the diagnostic accuracy for each study, are modeled as random effects by means of independent, variances of normal distributions: $\theta_i \sim N(\Theta, \sigma_\theta^2)$ and $\alpha_i \sim N(\Lambda, \sigma_\alpha^2)$, respectively. The modeling also includes a parameter of shape or scale β that allows the asymmetry in the SROC curve when admitting that the precision varies with the implicit threshold.

Therefore, the SROC curve is symmetric if $\beta = 0$ or asymmetric if $\beta \neq 0$. Each study provides a single point in the ROC space, and therefore, the estimation of β requires information from all the studies included in the meta-analysis. Thus, β is modeled as a fixed effect. The HSROC model has the following five parameters: Λ, Θ, β, σ_α^2 y σ_θ^2; the model is reduced to a fixed-effect model if: $\sigma_\alpha^2 = \sigma_\theta^2 = 0$. Other specifications for SROC curves based on Bivariate model functions have been proposed [5, 11], in this work we will focus our interest on the Rutter and Gatsonis model.

2 Methodology

A simulation study was carried out to compare the HSROC model with various simplifications (by eliminating the parameters of the model). We have chosen the HSROC model because it has greater flexibility when choosing parameters [12]. The specifications of the scenarios were designed to reproduce realistic situations found in the meta-analysis of diagnostic accuracy studies. The effect of these factors was investigated: (1) number of studies; (2) prevalence of the disease; (3) variability between the studies in accuracy and threshold; and (4) asymmetry in the SROC curve. We only investigate methods that use a binomial distribution.

2.1 *Generation of the Simulated Data*

Meta-analyzes with different numbers of studies were investigated randomly (N = 5, 10, 20, 35). The size of a study in each meta-analysis, n_j, was randomly sampled from a uniform distribution, $U(200; 2000)$; Given an underlying p prevalence, individuals within each study were randomly classified as sick or undiagnosed, and a continuous test result value, x, was assigned that was sampled randomly [13]. To determine the corresponding tetrachoric tables, we used the HSROC package of the statistical program R [14], by using the simdata function (see Fig. 2). To create the $2x2$ *table* for each study, the individuals were classified as true positives, false negatives, false positives or true negatives based on the result of the test and the state of

Table 2 Structure of the models adjusted to the simulated data

Num.	Models	Λ	Θ	β	σ_α^2	σ_θ^2
1	Complete model HSROC	√	√	√	√	√
2	Symmetrical HSROC model	√	√		√	√
3	HSROC model with fixed thresholds	√	√	√	√	
4	HSROC model with fixed precision	√	√	√		√
5	HSROC model with fixed threshold and accuracy	√	√	√		

the disease. The prevalences of the studies were adjusted by a uniform distribution with parameters 0 and 0.25, that is, the prevalence of these studies took values less than or equal to 25%.

We generated 23110 studies in 1200 independent meta-analysis datasets to allow an accurate estimation of model performance, even if a large proportion of models do not converge. In order to execute the simdata function, it is necessary to specify the input parameters of the function, that is, it is necessary to choose the a priori distributions of the simdata function, the initial parametrization was performed using the following initial conditions [15]:

$$Dataset \begin{cases} N \sim Uniform\,[5, 35] \\ n \sim Uniform\,[200, 2000] \\ p \sim Uniform\,[0.0001, 0.2499] \\ \beta \sim Uniform\,[-0.75, 0.75] \\ \Lambda \sim Uniform\,[-0.75, 0.75] \\ \sigma_\alpha^2 \sim Uniform\,[0.0001, 2] \\ \Theta \sim Uniform\,[-1.5; 1.5] \\ \sigma_\theta^2 \sim Uniform\,[0.0001, 2] \end{cases}$$

Additionally, in each generated meta-analysis, a covariate with categories from 1 to 3 was included, through a uniform distribution $U(1, 3)$.

2.2 *Hierarchical Models Adjusted to the Simulated Data*

In the later part we refer to an HSROC model that contains five parameters. The following five models were adjusted to the 1200 sets of data generated. From Table 2, it follows that for an HSROC model with precision and fixed thresholds (model 5), it is only necessary to identify a priori the distributions for, Λ, Θ, and β. Similarly, the remaining four models are understood.

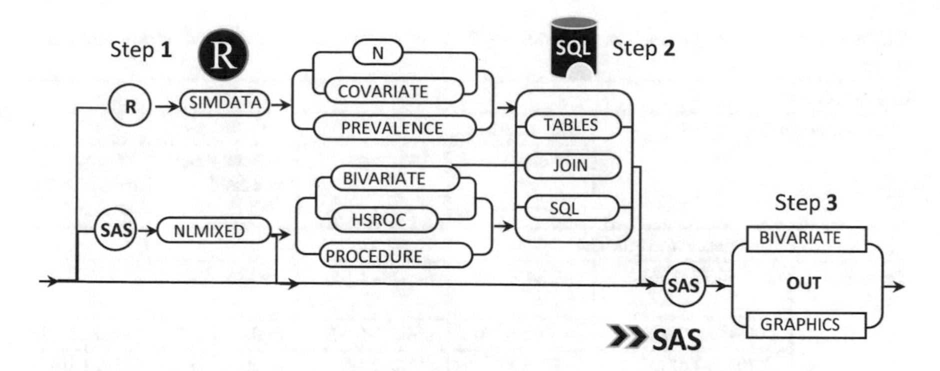

Fig. 2 General scheme of work and software required for the simulation of the data set (dataset), storage in database [16] and statistical analysis for the presentation of results [17]

Note that due to the mathematical relationship between the bivariate model and HSROC, it is possible to find the estimates of the five parameters of the HSROC model by means of the reitsma function of the R mada package [6].

3 Results

The results of the simulated data (Table 3) show the different models that were obtained by varying the hyperparameters of the complete hierarchical model.

Each category of the model is compared according to the adjustment criteria (−2LogLikelihood, AIC, AICC, and BIC) with their respective confidence intervals.

For example, we see that the best model under prevalence conditions between 0.01 and 25% and the selection based on the lowest information criterion of Akaike "AIC" is Model 2, in which case it corresponds to a symmetric HSROC model with precision and threshold fixed.

Table 4 shows the estimates of the parameters of the bivariate model through the execution of the NLMIXED procedure and the incorporation or not of a covariate.

It is important to emphasize that the incorporation of a covariate to the model provides better adjustment to the modeling (see −2Log Likelihood).

The convergence problems of the HSROC models were presented with greater recurrence when: (1) the estimated variance of the parameters $(\sigma_\alpha^2, \sigma_\theta^2)$ were very close to zero; (2) when the relation that the variance of the parameter α is twice or more the variance of the parameter θ, i.e., $(\sigma_\alpha^2/\sigma_\theta^2 > 2)$ is satisfied; (3) when the variance of the random effects of the threshold and the precision are very small or close to zero, this is understandable since the HSROC model was designed to model the random effects.

Table 3 Adjustment of the HSROC models, with parameters in each model and adjustment measures

HSROC Model					
Without covariate		With covariate			
		Threshold	Accuracy	Accuracy and threshold	Accuracy and threshold and shape
	Confidence intervals 95%				
Parameter estimates	Λ: (−0.37, 0.26)	(−0.34, 0.29)	(−0.29, 0.35)	(−0.33, 0.31)	(−0.33, 0.30)
	θ: (−0.31, 0.03)	(−0.32, 0.03)	(−0.28, 0.07)	(−0.30, 0.04)	(−0.30, 0.05)
	β: (0.99, 1.17)	(0.99, 1.17)	(1.00, 1.18)	(0.98, 1.17)	(0.97, 1.16)
	σ_α^2: (2.49, 3.52)	(2.58, 3.90)	(2.40, 3.42)	(2.50, 3.53)	(2.49, 3.52)
	σ_θ^2: (1.32, 1.68)	(1.43, 1.83)	(1.35, 1.75)	(1.38, 1.79)	(1.34, 1.73)
Goodness of fit statistics	−2logl: (1027, 1161)	(1012, 1145)	(1022, 1155)	(1014, 1147)	(1019, 1153)
	AIC: (1037, 1171)	(1024, 1157)	(1034, 1167)	(1028, 1161)	(1035, 1169)
	AICC: (1040, 1174)	(1029, 1161)	(1039, 1171)	(1035, 1168)	(1048, 1182)
	BIC: (1041, 1175)	(1029, 1162)	(1039, 1172)	(1033, 1166)	(1042, 1176)

Table 4 Adjustment of the bivariate model, bivariate model with and without covariate

Bivariate model		
	Without covariate	With covariate
	Confidence intervals 95%	
Parameter estimates	μ_A: (−4.48, −3.56)	(−4.11, −3.27)
	μ_B: (−0.12, −0.11)	(−0.30, 0.14)
	σ_A^2: (−0.16, −0.07)	(−0.14, 0.10)
	σ_B^2: (5.25, −12.85)	(2.58, 3.90)
	σ_{AB}: (7.05, −8.39)	(6.52, 7.53)
Fit statistics	−2logl: (372.23, 392.42)	(370.29, 390.5)
	AIC: (382.23, 402.42)	(384.29, 404.5)
	AICC: (385.07, 405.03)	(390.99, 410.57)
	BIC: (386.52, 407)	(390.3, 410.9)

4 Discussion

The categories of hierarchical models obtained by varying the hyperparameters directly model the precision and the threshold in the generalized mixed nonlinear model [9], with the simulation of the different scenarios and using the available software, we investigate situations extremes of the hierarchical HSROC model to understand its complex execution in meta-analysis for diagnostic accuracy tests. Convergence problems arose when the variance of one of the random effects is close to zero. This is particularly a problem for the parameterization of the bivariate model, where an examination of the scatter plot can help to identify a strong heterogeneity in sensitivity. We recommend a bivariate approach when the heterogeneity between the included studies is moderate, and it is necessary to estimate a summary measure for the sensitivity and specificity, but if the degree of heterogeneity is significant, we recommend an HSROC model if the estimation of an SROC curve is necessary with their respective variations.

Appendix

Availability and requirements for R Program
Project name: BivariateHSROC_Simulation
Project homepage: https://sourceforge.net/projects/metahi/
File: sourceCodeR.txt
Operating system(s): Microsoft Windows, Linux and Mac
Programming language: R, **License:** Open Source and free

Availability and requirements for SAS Program
Project name: BivariateHSROC_Simulation
Project homepage: https://sourceforge.net/projects/metahi/
File: sourceCodeSAS.txt
Operating system(s): Microsoft Windows, Linux and Mac
Programming language: SAS, **License:** Open Source and free

References

1. Midgette, A.S., Stukel, T.A., Littenberg, B.: A meta-analytic method for summarizing diagnostic test performances: receiver-operating-characteristic-summary point estimates. Med. Decis. Mak. **13**(3), 253–257 (1993)
2. Moses, L.E., Shapiro, D., Littenberg, B.: Combining independent studies of a diagnostic test into a summary ROC curve: data-analytic approaches and some additional considerations. Stat. Med. **12**(14), 1293–1316 (1993)
3. Reitsma, J.B., Glas, A.S., Rutjes, A.W., Scholten, R.J., Bossuyt, P.M., Zwinderman, A.H.: Bivariate analysis of sensitivity and specificity produces informative summary measures in diagnostic reviews. J. Clin. Epidemiol. **58**(10), 982–990 (2005)
4. Rutter, C.M., Gatsonis, C.A.: A hierarchical regression approach to meta-analysis of diagnostic test accuracy evaluations. Stat. Med. **20**(19), 2865–2884 (2001)
5. Arends, L.R., Hamza, T.H., Van Houwelingen, J.C., Heijenbrok-Kal, M.H., Hunink, M.G.M., Stijnen, T.: Bivariate random effects meta-analysis of ROC curves. Med. Decis. Mak. **28**(5), 621–638 (2008)
6. Harbord, R.M., Deeks, J.J., Egger, M., Whiting, P., Sterne, J.A.: A unification of models for meta-analysis of diagnostic accuracy studies. Biostatistics **8**(2), 239–251 (2006)

7. Kotz, S., Balakrishnan, N., Johnson, N.L.: Bivariate and trivariate normal distributions. Contin. Multivar. Distrib. **1**, 251–348 (2000)
8. Van Houwelingen, H.C., Arends, L.R., Stijnen, T.: Advanced methods in meta-analysis: multivariate approach and meta-regression. Stat. Med. **21**(4), 589–624 (2002)
9. Chu, H., Guo, H., Zhou, Y.: Bivariate random effects meta-analysis of diagnostic studies using generalized linear mixed models. Med. Decis. Mak. **30**(4), 499–508 (2010)
10. Macaskill, P.: Empirical Bayes estimates generated in a hierarchical summary ROC analysis agreed closely with those of a full Bayesian analysis. J. Clin. Epidemiol. **57**(9), 925–932 (2004)
11. Chappell, F.M., Raab, G.M., Wardlaw, J.M.: When are summary ROC curves appropriate for diagnostic meta-analyses? Stat. Med. **28**(21), 2653–2668 (2009)
12. Kriston, L., Hölzel, L., Weiser, A.K., Berner, M.M., Härter, M.: Meta-analysis: are 3 questions enough to detect unhealthy alcohol use? Ann. Intern. Med. **149**(12), 879–888 (2008)
13. Bachmann, L.M., Puhan, M.A., Ter Riet, G., Bossuyt, P.M.: Sample sizes of studies on diagnostic accuracy: literature survey. Bmj **332**(7550), 1127–1129 (2006)
14. Doebler, P., Holling, H.: Meta-analysis of diagnostic accuracy with mada. Reterieved at: https://cran.rproject.org/web/packages/mada/vignettes/mada.pdf (2015)
15. Dendukuri, N., Schiller, I., Joseph, L., Pai, M.: Bayesian meta-analysis of the accuracy of a test for tuberculous pleuritis in the absence of a gold standard reference. Biometrics **68**(4), 1285–1293 (2012)
16. Grycuk, R., Gabryel, M., Scherer, R., Voloshynovskiy, S.: Multi-layer architecture for storing visual data based on WCF and microsoft SQL server database. In: International Conference on Artificial Intelligence and Soft Computing, pp. 715–726. Springer, Cham (2015)
17. Sheu, C.F., Chen, C.T., Su, Y.H., Wang, W.C.: Using SAS PROC NLMIXED to fit item response theory models. Behav. Res. Methods **37**(2), 202–218 (2005)

Compound Dirichlet Processes

Arrigo Coen and Beatriz Godínez-Chaparro

Abstract The compound Poisson process and the Dirichlet process are the pillar structures of renewal theory and Bayesian nonparametric theory, respectively. Both processes have many useful extensions to fulfill the practitioners' needs to model the particularities of data structures. Accordingly, in this contribution, we join their primal ideas to construct the compound Dirichlet process and the compound Dirichlet process mixture. As a consequence, these new processes have a rich structure to model the time occurrence among events, with also a flexible structure on the arrival variables. These models have a direct Bayesian interpretation of their posterior estimators and are easy to implement. We obtain expressions of posterior distribution, nonconditional distribution, and expected values. In particular, to find these formulas, we analyze sums of random variables with Dirichlet process priors. We assess our approach by applying our model on a real data example of a contagious zoonotic disease.

Keywords Bayesian nonparametrics · Renewal theory · Compound poisson process · Dirichlet process · Random sums

1 Introduction

In this contribution, we present two continuous time processes that are probabilistically constructed through random sums, using the framework of Bayesian nonparametric models. As a consequence of their construction, these processes can be

A. Coen (✉)
Facultad de Ciencias, Departamento de Matemáticas, Universidad Nacional Autónoma de México, Apartado Postal 20-726, 01000 Mexico, CDMX, Mexico
e-mail: coen@ciencias.unam.mx

B. Godínez-Chaparro
Departamento de Sistemas Biológicos, División de Ciencias Biológicas y de la Salud, Universidad Autónoma Metropolitana–Xochimilco, Mexico City, Mexico
e-mail: bgodinez@correo.xoc.uam.mx

I. Antoniano-Villalobos et al. (eds.), *Selected Contributions on Statistics and Data Science in Latin America*, Springer Proceedings in Mathematics & Statistics 301,
https://doi.org/10.1007/978-3-030-31551-1_4

used to model renewal phenomena. Examples of applied Bayesian nonparametric models to analyze renewal theory phenomena are presented in [6, 17, 39]. One of the principal reasons to combine these methodologies is the fact that in many cases the renewal phenomena have complex random structures. Therefore, for these types of analyses, it could be better to let the data speak by itself. By using parameter-free models, important hidden structures unveil, whereas a parametric model may conceal them. Although the combination of these branches is not new, the use of the Dirichlet process that is here presented is a novel technique.

For many applied statisticians random sums models are everyday tools. An advantage of these models is that they allow us to examine the data as the contribution of simpler parts, which improve calculations and predictions. To choose a random sum model, there are three key probability concepts to have in mind: (1) the law governing the number of terms to add; (2) the dependence among the terms; and (3) the interactions between (1) and (2). For instance, in [35] these concepts are applied to model the behavior of insurance claims by taking into account: (1) how many insurance claims are received in a fixed period of time; (2) the dependence of claims sizes; and (3) the connections between the number of claims and their sizes (see also [8, 18]). Other fields where random sum models are currently applied are Multivariate analysis to model daily stock values [30], Bayesian nonparametric theory to estimate the total number of species in a population [41], and Finance to estimate the skewed behavior of a time series [29].

The classical theory of renewal processes focuses on the analysis of counting process where the interarrival times are independent and identically distributed (i.i.d.). The most remarkable example of renewal process is the Poisson process, whose interarrival times are i.i.d. exponential variables [22]. By allowing some interaction among the variables, this model has been generalized to resemble more intricate phenomena. Examples of these generalizations are the Cox process, the nonhomogeneous Poisson process, and the Markov and semi-Markov renewal models. A thorough analysis of these models is presented in [24].

To define our model, we use one of the most influential Bayesian nonparametric structures, the Dirichlet process (DP) prior [15]. The DP effectiveness is attested by its successful application in many statistical analyses. As pointed out by Ferguson in [15], two desirable properties of a prior distribution for nonparametric problems are: a large support and a manageable posterior distribution. The DP prior handles both properties in a remarkable manner, with a clear interpretation of its parameters. Moreover, the many representations of the Dirichlet process give rise to diverse important Bayesian nonparametric structures: neutral to the right processes [10], normalized log-Gaussian processes [25], stick breaking priors [21, 26], species sampling models [32], Poisson–Kingman models [33], and normalized random measures with independent increments [34], to mention a few. Each of these models generalizes an aspect of the Dirichlet process in some direction, thus obtaining more modeling flexibility with respect to some specific features of the data.

In this study, we are applying the Dirichlet process as a mechanism to control the probability structure of a random sum stochastic process. Under this framework, we inherit the flexibility of the DP to resemble the data behavior and have a broad spectrum of probability structures to establish as prior beliefs. Also, we gain an interpretation of the clustering structure of the renewals and an efficient posterior simulation algorithm. In fact, these models allow us to analyze the cluster behavior of the time and space components, induced by discrete random measures.

2 Compound Dirichlet Processes

In this section, we define the stochastic structure of the compound Dirichlet process and the compound Dirichlet process mixture, and show some of their appealing modeling properties. These processes could be applied to phenomena where the stochastic-time component defines the arrivals of random variables. Under this framework, we settle a dependence structure among arrivals and another among the events of the arrivals, keeping independence between the two.

Let us first consider a sequence of positive random variables $\{T_j\}_{j=1}^{\infty}$, and define its renewal process $\{N_t\}_{t\in\mathbb{R}_+}$ as

$$N_t = \sup\{j \in \mathbb{N} : T_1 + T_2 + \cdots + T_j < t\}, \quad t \in \mathbb{R}_+,$$

where the random variables T_j are interpreted as the interarrival times between events of the phenomenon of study. Then, N_t is the number of events that take place before time t. The general theory of exchangeable renewal models is studied in [7], however, here we analyze the particular implications of the DP prior framework. To this end, similar to the ideas of a compound Poisson process, we focus our analysis on the random process $\{S_t\}_{t\in\mathbb{R}_+}$ given by

$$S_t = \sum_{i=1}^{N_t} X_i, \qquad t \in \mathbb{R}_+, \tag{1}$$

where $\{X_i\}_{i=1}^{\infty}$ is independent of $\{T_j\}_{j=1}^{\infty}$. In this construction, we will place two exchangeable structures: one over the events $\{X_i\}_{i=1}^{\infty}$ and one on their interarrival times $\{T_j\}_{j=1}^{\infty}$. The advantage of assuming this symmetric structure lies in the fact that with it we could model various dependence behaviors and, at the same time, allows the analysis of cluster formations among variables [1]. To define these dependent structures, we will use Dirichlet process priors. The DP prior model is defined as

$$\begin{aligned} X_i \mid G &\sim G, \\ G &\sim \mathrm{DP}(\alpha, G_0), \end{aligned} \tag{2}$$

where $\mathrm{DP}(\alpha, G_0)$ denotes a Dirichlet process with precision parameter $\alpha > 0$ and base distribution G_0. The DP random measure G is defined in [15] by the distributional property

$$(G(A_1), \dots, G(A_k)) \sim \text{Dir}(\alpha G_0(A_1), \dots, \alpha G_0(A_k)),$$

for all measurable partition $(A_1, \dots, A_k)$ of the sample space of G_0, where $\text{Dir}(a_1, \dots, a_k)$ denotes the Dirichlet distribution of k-dimension with parameter $(a_1, \dots, a_k)$. An implication of these assumptions is that the joint distribution of $(X_1, \dots, X_n)$ can be factorized using the generalized Pólya urn scheme [3], i.e., for any $n > 1$,

$$X_n | X_{n-1}, X_{n-2}, \dots, X_1 \sim \frac{\alpha}{\alpha + n - 1} G_0 + \frac{1}{\alpha + n - 1} \sum_{i=1}^{n-1} \delta_{X_i}, \tag{3}$$

where δ_x denotes de Dirac measure at x. This last expression could be interpreted as X_n given $X_{n-1}, X_{n-2}, \dots, X_1$ has probability $\frac{\alpha}{\alpha+n-1}$ of being a new G_0-distributed random variable independent of the past values and probability $\frac{n-1}{\alpha+n-1}$ to repeat a previously seen value. This also implies that the random variables $\{X_i\}_{i=1}^{\infty}$ are exchangeable, meaning that the joint distribution of $(X_1, \dots, X_n)$ is equal to the distribution of $(X_{\pi_1}, \dots, X_{\pi_n})$, for any permutation π of $\{1, \dots, n\}$ [1]. It follows that the variables X_i are conditionally independent and identically distributed G_0, with constant correlation given by

$$\text{Corr}(X_i, X_j) = \frac{1}{\alpha + 1}, \qquad i, j \in \mathbb{N}.$$

It is important to take into consideration the discreteness of the distributions sampled from the Dirichlet process. Many works overcome this difficulty by using a DP as a prior over the distribution of an extra layer of parameters [12, 16, 27]. In fact, in many cases, these parameters help to make the description simpler and have a direct interpretation. These models are known as the Dirichlet process mixtures models, and they are defined by the structure

$$\begin{aligned} X_i \mid \theta_i &\sim F_{\theta_i} \\ \theta_i \mid G &\sim G, \\ G &\sim \text{DP}(\alpha, G_0), \end{aligned}$$

where F_θ denotes a member of a fixed family of distributions parametrized by θ. Even thought, this last approach adds a hidden extra layer of parameters, there are many Gibbs sampling methods to confront this issue [21, 28, 31]. Furthermore, the discreteness of the random DP measures allows to study the clustering properties of the data [11, 13, 20]. Under this notation, we can establish a nonparametric structure on (1).

Definition 1 A continuous time stochastic process $\{S_t\}_{t \in \mathbb{R}_+}$ given by (1) is a compound Dirichlet process (CDP) if it follows the stochastic structure

$$T_j \mid G^T \sim G^T, \qquad X_i \mid G^X \sim G^X,$$
$$G^T \sim \text{DP}(\alpha^T, G_0^T), \qquad G^X \sim \text{DP}(\alpha^X, G_0^X),$$

where $\{X_i\}_{i=1}^{\infty}$ is independent of $\{T_j\}_{j=1}^{\infty}$. To simplify the notation, we use $S_t \sim \text{CDP}(\alpha^T, G_0^T, \alpha^X, G_0^X)$.

It is important to notice that, as in the DP framework, the CDP model also has a positive probability of repeating previously observed values. In the classical DP model, as $n \to \infty$ the expected number of distinct X_i terms in $\{X_1, \ldots, X_n\}$ grows as $\alpha \log n$ [23]; it is important to notice that this rate is smaller than n. Consequently, the CDP has a positive probability of repeat increments. In other words, there is a positive probability that the increment $S_{t_2} - S_{t_1}$ is equal to the increment $S_{t_4} - S_{t_3}$, for any positive real numbers $t_1 < t_2$ and $t_3 < t_4$. Nevertheless, it is important to notices that the rate of repeated values is even smaller than the one of the DP frameworks. The addition operation confers a decrease in the number of repeated values; selecting different adding terms gives an extra possibility of different total results. In order to diminish the problem of repeated values and to study the clustering structure of the random variables, we have the next definition.

Definition 2 A continuous time stochastic process $\{S_t\}_{t \in \mathbb{R}_+}$ given by (1) is a compound Dirichlet process mixture (CDPM) if it follows the stochastic structure

$$T_j \mid \theta_j^T \sim F_{\theta_j}^T \qquad X_i \mid \theta_i^X \sim F_{\theta_i}^X$$
$$\theta_j^T \mid G^T \sim G^T, \qquad \theta_i^X \mid G^X \sim G^X,$$
$$G^T \sim \text{DP}(\alpha^T, G_0^T), \qquad G^X \sim \text{DP}(\alpha^X, G_0^X),$$

where $\{X_i\}_{i=1}^{\infty}$ is independent of $\{T_j\}_{j=1}^{\infty}$. We use $S_t \sim \text{CDPM}(\alpha^T, G_0^T, F^T, \alpha^X, G_0^X, F^X)$ to denote this process, where F^T and F^X represent parametric families of distributions.

Under Definitions 1 and 2, we have a rich structure to consider the time evolution of the compound random variables. The next section presents the statistical implications of the CDP and CDPM models.

2.1 Some Properties of CDP and CDPM

Let us continue with some properties of the CDP and CDPM models. These results are presented under the CDP framework; however, their implications on the CDPM models are direct. The results are arranged in order to calculate, or at least approximate, the posterior distribution of S_t.

Theorem 1 *If* $S_t \sim \text{CDP}(\alpha^T, G_0^T, \alpha^X, G_0^X)$, *then for any* $t, s \in \mathbb{R}_+$

$$\mathbb{P}[S_t \le s] = \sum_{n=1}^{\infty} \sum_{v \in \Delta_n} p_v(n) \left(H_1^{*v_1} * H_2^{*v_1} * \cdots * H_n^{*v_n}\right)(s) \mathbb{P}[N_t = n], \quad (4)$$

where $\Delta_n = \left\{v = (v_1, \ldots, v_n) \in \mathbb{N}^n : \sum_{i=1}^n i v_i = n\right\}$,

$$p_v(n) = \frac{n!}{\alpha^X(\alpha^X + 1) \cdots (\alpha^X + n - 1)} \prod_{i=1}^{n} \frac{(\alpha^X)^{v_i}}{i^{v_i} v_i!},$$

$H_i^{*v_i}$ *is the* v_i*-convolution of the distribution* $H_i(\cdot) = G_0^X(\cdot/i)$*, and* $\left(H_1^{*v_1} * H_2^{*v_1} * \cdots * H_n^{*v_n}\right)(\cdot)$ *is the convolution of these convolutions.*

The proof of Theorem 1 is a direct consequence of the following lemma.

Lemma 1 *If* $X_i|G \sim G$ *and* $G \sim DP(\alpha, G_0)$*, we define* $\{S_n\}_{n=1}^{\infty}$ *by*

$$S_n = \sum_{i=1}^{n} X_i, \quad n \in \mathbb{N}.$$

Then

$$\mathbb{P}[S_n \le s] = \frac{n!}{\alpha(\alpha+1) \cdots (\alpha+n-1)} \sum_{v \in \Delta_n} \left(H_1^{*v_1} * H_2^{*v_1} * \cdots * H_n^{*v_n}\right)(s) \prod_{j=1}^{n} \frac{\alpha^{v_j}}{j^{v_j} v_j!},$$

for every $s \in \mathbb{R}_+$*, where* $H_i^{*v_i}$ *is the* v_i*-convolution of the distribution* $H_i(\cdot) = G_0^X(\cdot/i)$*. Moreover, if we define*

$$N_t^X = \sup\{n \in \mathbb{N} : X_1 + X_2 + \cdots + X_n < t\}, \quad t \in \mathbb{R}_+.$$

then $\mathbb{P}\left[N_t^X = n\right] = \mathbb{P}[S_n \le t]$ *for* $t > 0$.

Proof For the sake of completeness, we repeat the proof presented in [7] for this result. Let $V = (V_1, \ldots, V_n)$ be the random vector counting the repeated values in $(X_1, \ldots, X_n)$, under the following scheme: there are V_1 values that only repeat once, V_2 values that repeat twice, and so on. Then, conditioning on V the distribution of S_n can be written as

$$\mathbb{P}[S_n \le s] = \sum_{v \in \Delta_n} \mathbb{P}[X_1 + \cdots + X_n \le s | V = v] \mathbb{P}[V = v].$$

The conditional distribution of $X_1 + \cdots + X_n$ given $V = v$ is equal to the convolution of v_1 independent variables with distribution H_1, convolved with the convolution of v_2 variables distributed H_2, and so on. We condition on V because this eliminates the repeated values of $(X_1, \ldots, X_n)$, allowing us to consider the convolution of independent variables. Consequently, we define H_j because given the repeated values of

X_i we need to consider the probabilities $\mathbb{P}[jX \leq t]$, for $X \sim G_0$ and $j \in \mathbb{N}$. Thus, we obtain

$$\mathbb{P}[X_1 + \cdots + X_n \leq s | V = v] = \left(H_1^{*v_1} * H_2^{*v_1} * \cdots * H_n^{*v_n}\right)(s).$$

The probabilities of $\{V = v\}$ are given by the Ewens's sampling formula [14], as

$$\mathbb{P}[V = v] = \frac{n!}{\alpha(\alpha+1)\cdots(\alpha+n-1)} \prod_{j=1}^{n} \frac{\alpha^{v_j}}{j^{v_j} v_j!}, \tag{5}$$

by applying the generalized Pólya urn scheme (3) over the possible different values of $X_1, \ldots, X_n$. Finally, the equality $\mathbb{P}\left[N_t^X = n\right] = \mathbb{P}[S_n \leq t]$ is a direct consequence of the definition of N_t^X. □

According to the last results, the distribution of the CDP can be expressed as an infinite sum. Although we are not presenting directly the distributions of S_t and N_t, they can be expressed in terms of the distribution of S_n, using the second statement of Lemma 1. Since the Dirichlet process tends to concentrate most of its mass on a few atoms the convergence of the series of (4) is fast. This allows us to approximate the distribution of S_t in two ways. We can truncate the sum (4) to a finite fixed number of terms, or we can fix a quantity ε_0 to count only terms with $\mathbb{P}[N_t = n] > \varepsilon_0$. In both cases, we restrain the error of the approximation. Also, our computational experiments show that both approximations are stable.

Proposition 1 *Given $S_t \sim \mathrm{CDP}(\alpha^T, G_0^T, \alpha^X, G_0^X)$, let $\mu_{X,i} = \mathbb{E}\left[X_1^i\right]$ and $\mu_{N_t,i} = \mathbb{E}\left[N_t^i\right]$, for $i = 1, 2, 3$, then*

$$\mathbb{E}[S_t] = \mu_{N_t,1}\mu_{X,1}$$

$$\mathbb{E}\left[S_t^2\right] = \frac{\left(\mu_{N_t,2} - \mu_{N_t,1}\right)\left(\alpha\mu_{X,1}^2 + \mu_{X,2}\right)}{\alpha+1} + \mu_{N_t,1}\mu_{X,2}$$

$$\mathbb{E}\left[S_t^3\right] = \frac{\left(2\mu_{N_t,1} - 3\mu_{N_t,2} + \mu_{N_t,3}\right)\left(\alpha^2\mu_{X,1}^3 + 3\alpha\mu_{X,2}\mu_{X,1} + 2\mu_{X,3}\right)}{(\alpha+1)(\alpha+2)} + \frac{\left(\mu_{N_t,2} - \mu_{N_t,1}\right)\left(\alpha\mu_{X,1}\mu_{X,2} + \mu_{X,3}\right)}{\alpha+1} + \mu_{N_t,1}\mu_{X,3}$$

Proof The expression for $\mathbb{E}[S_t]$ follows conditioning on $\{N_t = n\}$ and using the lineality of the expectation operator. To obtain the expression for $\mathbb{E}\left[S_t^2\right]$, one must consider the possible repeated values of the exchangeable sequence $\{X_i\}$. From now, let us assume that G_0^X is a continuous distribution. Then, conditioning on $\{N_t = n\}$, we obtain

$$\begin{aligned}
\mathbb{E}\left[S_t^2|N_t=n\right] =& \mathbb{E}\left[\left(\sum_{i=1}^{n} X_i\right)^2\right] \\
=& \sum_{i=1}^{n}\mathbb{E}\left[X_i^2\right] + \sum_{1\le i<j\le n}\mathbb{E}\left[X_iX_j\right] \\
=& n\mu_{X,2} + n(n+1)\mathbb{E}\left[X_1X_2\right] \\
=& n\mu_{X,2} + n(n+1)\left(\mu_{X,1}^2\frac{\alpha}{\alpha+1} + \mu_{X,2}\frac{1}{\alpha+1}\right).
\end{aligned}$$

The last equality follows from conditioning on $\{X_1 = X_2\}$, and knowing that $\mathbb{P}\left[X_1 = X_2\right] = \frac{\alpha}{\alpha+1}$. This last expression gives the result for $\mathbb{E}\left[S_t^2\right]$. Likewise, the result for $\mathbb{E}\left[S_t^3\right]$ is obtained by conditioning on the possible repetitions of $\{X_1, X_2, X_3\}$, and applying $\mathbb{P}\left[X_1 = X_2 = X_3\right] = \frac{1}{\alpha+1}\frac{2}{\alpha+2}$, $\mathbb{P}\left[X_1 = X_2 \neq X_3\right] = \frac{1}{\alpha+1}\frac{\alpha}{\alpha+2}$ and $\mathbb{P}\left[X_1 \neq X_2 \neq X_3 \neq X_1\right] = \frac{\alpha}{\alpha+1}\frac{\alpha}{\alpha+2}$. Finally, in the discrete case, we only need to ensure that we are conditioning only on cases when the variables are equal as a consequence of the Pólya urn's repetitions, and the formulas follow. □

Proposition 1 *Under the notation of Lemma 1, the moment generator function of S_n is given by*

$$M_{S_n}(t) = \frac{n!}{\alpha(\alpha+1)\cdots(\alpha+n-1)}\sum_{v\in\Delta_n}\prod_{j=1}^{n}\frac{1}{v_j!}\left[\frac{\alpha M_X(tj)}{j}\right]^{v_j}, \tag{6}$$

where M_X denotes the generating generator function of X_1.

Proof Conditioning over the possible partitions, we obtain

$$\begin{aligned}
M_{S_n}(t) &= \sum_{v\in\Delta_n}\mathbb{P}\left[V=v\right]\mathbb{E}\left[e^{tS_n}|V=v\right] \\
&= \sum_{v\in\Delta_n}\mathbb{P}\left[V=v\right]\prod_{j=1}^{n}M_X(tj)^{v_j},
\end{aligned}$$

where the last equality follows since the expected value of e^{tS_n} conditioned on $\{V = v\}$ is the product of independent random variables equal in distribution to e^{tjX_1}, each repeated v_j times for $j = 1, \ldots, n$. This gives (6) when applying (5). □

To see an application of (6), let us consider the Gaussian distribution case. For this base distribution, we obtain

$$M_{S_n}(t) = \frac{n!}{\alpha(\alpha+1)\cdots(\alpha+n-1)} \sum_{v\in\Delta_n} \prod_{j=1}^{n} \frac{1}{v_j!}\frac{\alpha^{v_j}}{j^{v_j}} e^{t(\mu j v_j)+t^2(\sigma^2 j^2 v_j)/2}$$

$$= \frac{n!}{\alpha(\alpha+1)\cdots(\alpha+n-1)} \sum_{v\in\Delta_n} e^{t(n\mu)+t^2(\sigma^2 \sum_{j=1}^{n} j^2 v_j)/2} \prod_{j=1}^{n} \frac{1}{v_j!}\frac{\alpha^{v_j}}{j^{v_j}},$$

thus, we obtain that the sum of variables with DP prior and Gaussian base measure is a mixture of Gaussian random variables. The next result shows that the CDP is a conjugate model.

Proposition 2 *If* $(X_1, T_1), \ldots, (X_n, T_n)$ *is a random sample of* $S_t \sim \mathrm{CDP}(\alpha^T, G_0^T, \alpha^X, G_0^X)$*, then*

$$S_t \mid (X_1, T_1), \ldots, (X_n, T_n)$$
$$\sim \mathrm{CDP}\left(\alpha^T + n, \frac{\alpha}{\alpha+n} G_0^T + \frac{\alpha}{\alpha+n}\sum_{j=1}^{n} \delta_{T_j}, \alpha^X + n, \frac{\alpha}{\alpha+n} G_0^X + \frac{\alpha}{\alpha+n}\sum_{i=1}^{n} \delta_{X_i}\right). \quad (7)$$

Proof The proof is immediate by applying the conjugate property of the Dirichlet process prior and the independence between $\{T_j\}_{j=1}^{\infty}$ and $\{X_i\}_{i=1}^{\infty}$. □

2.2 Two Examples of Flexible Base Measures for CDP and CDPM

Let us continue by presenting two examples of families of distributions which when used as the base measure G_0^X simplify the convolution of (4). These families are the Gaussian and the phase-type. It is important to notice that both families have wide support, which allows their use to approximate other distributions. First, in the case of Gaussian distributions given by $G_0^X = \mathrm{N}(\mu, \sigma^2)$, we obtain that $H_j = \mathrm{N}(\mu j, \sigma^2 j^2)$, and so

$$H_1^{*v_1} * H_2^{*v_1} * \cdots * H_n^{*v_n} = \mathrm{N}\left(\mu n, \sigma^2 \sum_{j=1}^{n} j^2 v_j\right).$$

This implies that the density of S_n is given by

$$f_{S_n}(t) = \frac{n!}{\alpha(\alpha+1)\cdots(\alpha+n-1)} \sum_{v\in\Delta_n} \frac{e^{-(t-\mu n)^2/2\sigma^2 \sum_{j=1}^{n} j^2 v_j}}{\sqrt{2\pi\sigma^2 \sum_{j=1}^{n} j^2 v_j}} \prod_{j=1}^{n} \frac{\alpha^{v_j}}{j^{v_j} v_j!}.$$

Rates of the convergence of Gaussian mixtures to the true underlying distribution are presented in [19, 36]. As a consequence of this convergence, we could use the Gaussian model in cases with poor prior information.

For the second example, we present the analytic expression for f_{S_n} in the case of phase-type distributions. An excellent account of the theory of phase-type and matrix-exponential distributions is presented in [4]. An important property of this family is that it is dense on the set of positive random variables; i.e., any positive random variable can be arbitrarily approximated by a phase-type distribution. We denote by $U \sim \mathrm{PH}_p(\pi, \mathbf{T})$, a random variable with phase-type density given by

$$f(u) = \pi e^{\mathbf{T}u}\mathbf{t},$$

where $\pi = (\pi_1, \pi_2, \ldots, \pi_p)$ is a probability row vector, $\mathbf{T}$ a subgenerator matrix of dimension p, and $\mathbf{t} = -\mathbf{T}\mathbf{1}$, with $\mathbf{1}$ the vertical vector of ones of length p. Then, under this notation, if $G_0^X = \mathrm{PH}_p(\pi, \mathbf{T})$, we obtain that $H_j = \mathrm{PH}_p(\pi, \mathbf{T}/j)$. By applying the convolution property of phase-type variables:

$$Z_1 + Z_2 \sim \mathrm{PH}_{p_1+p_2}\left((\pi_1, 0), \begin{bmatrix} \mathbf{T}_1 & \mathbf{t}_1\pi_2 \\ 0 & \mathbf{T}_2 \end{bmatrix}\right),$$

for $Z_1 \sim \mathrm{PH}_{p_1}(\pi_1, \mathbf{T}_1)$ and $Z_2 \sim \mathrm{PH}_{p_2}(\pi_2, \mathbf{T}_2)$, with $\mathbf{t}_1 = -\mathbf{T}_1\mathbf{1}$. Thus,

$$H_j^{*v_j} = \mathrm{PH}_{pv_j}((\pi, 0, \ldots, 0), \mathbf{T}(j, v_j)),$$

where $\mathbf{T}(j, v_j)$ is a matrix of dimension $pv_j \times pv_j$, given by

$$\mathbf{T}(j, v_j) = \frac{1}{j}\begin{bmatrix} \mathbf{T} & -\mathbf{T}\mathbf{1}\pi & 0 & \ldots & 0 \\ 0 & \mathbf{T} & -\mathbf{T}\mathbf{1}\pi & \ldots & 0 \\ 0 & 0 & \mathbf{T} & \ldots & 0 \\ 0 & 0 & 0 & \ldots & 0 \\ \vdots & \vdots & \vdots & \ddots & \vdots \\ 0 & 0 & 0 & 0 & \mathbf{T} \end{bmatrix}.$$

This implies

$$H_1^{*v_1} * H_2^{*v_1} * \cdots * H_n^{*v_n} = \mathrm{PH}_{p\sum v_j}\left((\pi, 0, \ldots, 0), \mathbf{T}^v\right),$$

where $\mathbf{T}^v$ is a matrix of dimension $p\sum v_j \times p\sum v_j$, given by

$$\mathbf{T}^v = \begin{bmatrix} \mathbf{T}(1, v_1) & -\mathbf{T}\mathbf{1}\pi & 0 & 0 & \ldots & 0 \\ 0 & \mathbf{T}(2, v_2) & -\mathbf{T}\mathbf{1}\pi/2 & 0 & \ldots & 0 \\ 0 & 0 & \mathbf{T}(3, v_3) & -\mathbf{T}\mathbf{1}\pi/3 & \ldots & 0 \\ 0 & 0 & 0 & \mathbf{T}(4, v_4) & \ldots & 0 \\ \vdots & \vdots & \vdots & \vdots & \ddots & \vdots \\ 0 & 0 & 0 & 0 & 0 & \mathbf{T}(n, v_n) \end{bmatrix}.$$

We eliminate from $\mathbf{T}^{v}$ the rows where $v_j = 0$. This implies that the density of S_n is given by

$$f_{S_n}(u) = \frac{n!}{\alpha(\alpha+1)\cdots(\alpha+n-1)} \sum_{v\in\Delta_n} \pi e^{\mathbf{T}^{v}u}\mathbf{t}^{v} \prod_{j=1}^{n} \frac{\alpha^{v_j}}{j^{v_j} v_j!},$$

where $\mathbf{t}^{v} = -\mathbf{T}^{v}\mathbf{1}$. Thus, the sum of variables with DP prior and phase-type base measure is distributed as a mixture of phase-type random variables.

3 An Application to Rabies in Dogs

Rabies is one of the most severe zoonotic diseases. It is caused by a rhabdovirus in the genus Lyssavirus and infects many mammalian species. It can be transmitted through infected saliva, and it is almost fatal following the onset of clinical symptoms [37]. In up to 99% of cases, domestic dogs are responsible for the rabies virus transmission to humans. In Africa, an estimated 21,476 human deaths occur each year due to dog-mediated rabies, which is 36.4% of the global rabies-related human deaths [38]. To have effective intervention against zoonotic infections, it is important to recognize whether infected individuals stem from a single outbreak sustained by local transmission, or from repeated introductions [9, 40].

Some probability models commonly applied to model epidemiological contagion are coupling structures and random graphs. Likewise, the EM-algorithm and MCMC methods are frequently used to obtain predictions and confidence intervals. For a recent account of the theory, we refer the reader to [2, 5]. These models have dependence structures to represent the infectious rate as a function of infected individuals in the vicinity. Another important quality of these models is to admit censored data, since often the epidemic process is only partly observed. These two properties are also found in the CDPM model. The spatial vicinity is handled directly by the posterior distribution; areas where the cases of the disease are found have a bigger probability of new cases. Censored data can be managed both for censoring in time or in the space.

In [9], 151 cases of rabies in dogs reported in Bangui, the capital of the Central African Republic between the January 6, 2003 and the March 6, 2012, are analyzed. The data include information on report date, spatial coordinates, and genetic sequence of the isolated virus. The data are available in R the package `outbreaks`. The authors apply a clustering graph model for each component and extract the most connected dots by pruning. We study this data using the CDPM model under the following assumptions. To model the time component, we use an exponential mixture kernel with a $\text{Ga}(\alpha_0, \beta_0)$ base distribution, and to model the spatial component, we use a multivariate Gaussian mixture kernel with the natural choice of priors for the mean and covariance (Gaussian-inverse-Wishart distribution). The prior distribution of the mean μ

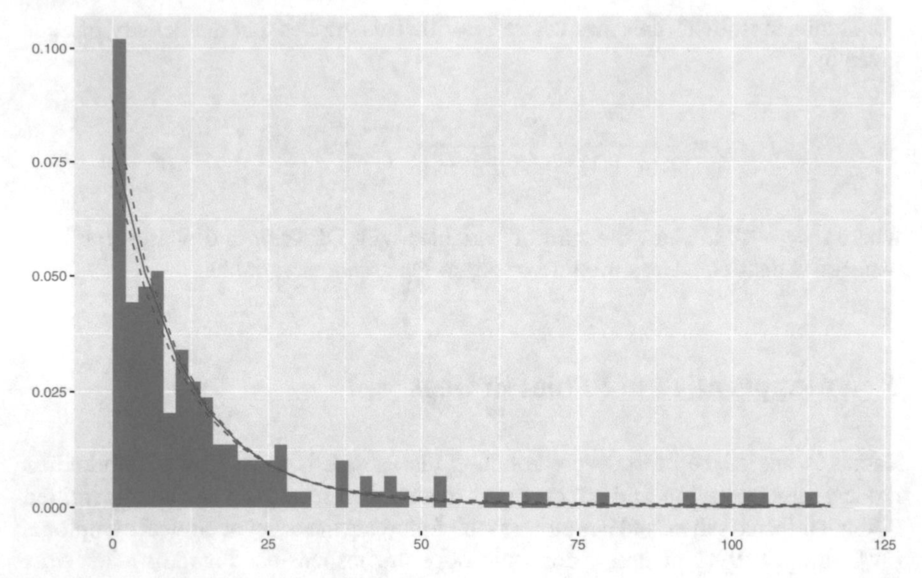

Fig. 1 Visualization of the density estimation of the rabies data using the CDPM model with respect to the time component, with an exponential mixture kernel and Ga(1, 8) base distribution. The x-axis represents the number of days between reports and the y-axis represents their frequency. In this figure is compared the data histogram against the predictive density estimator with a 95% confidence interval

$$(\mu|\Sigma, \xi, \rho) \sim \mathrm{N}(\xi, \rho\Sigma)$$

and the prior distribution of Σ^{-1} is Wishart:

$$(\Sigma^{-1}|\beta, W) \sim \mathrm{W}(\beta, (\beta W)^{-1}) = \frac{(|W|(\beta/2)^D)^{\beta/2}}{\Gamma_D(\beta/2)}|\Sigma|^{(\beta-D-1)/2} \exp\left(-\frac{\beta}{2}\mathrm{tr}(\Sigma W)\right),$$

where d is the dimension and $\Gamma_D(z) = \pi^{D(D-1)/4} \prod_{d=1}^{D} \Gamma\,(z + (d - D)/2)$. The joint distribution of μ and Σ is the Gaussian-inverse-Wishart distribution denoted as

$$(\mu, \Sigma) \sim \mathrm{NW}^{-1}(\xi, \rho, \beta, \beta W).$$

To fit these Dirichlet process mixtures, we use the Gibbs sampling methodology.

Figure 1 shows that the model is able to capture the density pattern of the time component. In this figure, the data histogram is compared against the predictive distribution given the data, with a 95% confidence interval. As pointed out by [9], the dates of the reports are close. This is characterized by the appearance of only two mixture components in the posterior distributions. Figure 2 presents the spatial cluster behavior of the predictive distribution. Even though some spatial data is missing, we estimate the spatial data from the posterior; these missing values do not affect our

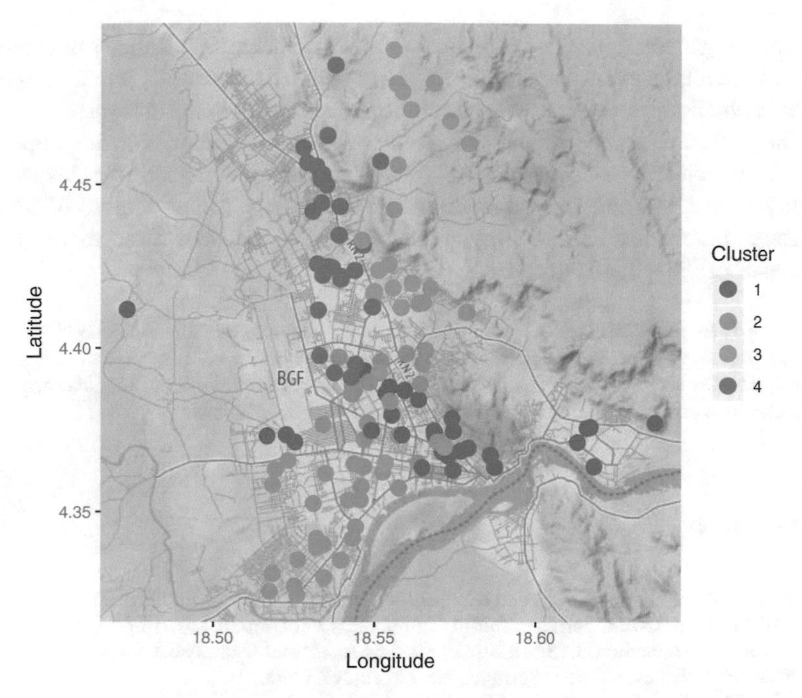

Fig. 2 Visualization of spatial clusters of the rabies data using the CDPM model, with a Gaussian mixture kernel and $\text{NW}^{-1}(0, 1, 1, I)$ base distribution

analysis since we are assuming independence between spatial and temporal variables. In comparison with the results of [9], we obtain almost the same cluster structure.

In this application, the value of applying the CDPM model is on the estimation of future rabies outbreaks. Under our framework, we obtain the complete probability structure of probable future contagion. We could answer many statistical questions through simulation using the posterior samples from the CDPM model. For instance, we can find the probability that in the next year, the number of cases doubles with respect to past year numbers. Likewise, we could obtain the spatial stochastic mobility of the disease, by locating the regions where the disease is more concentrated. Our model allows an early assessment of infectious disease outbreaks, which is fundamental to implementing timely and effective control measures.

4 Discussion

We have proposed a simple approach for statistical analysis of renewal phenomena, which combines ideas from renewal theory and Bayesian nonparametric theory. The model here presented is defined using two independent Dirichlet processes: one to model the time occurrence of events and another to model their spatial distribution.

The resulting methodology is not computationally demanding and allows us to predict relatively well the evolution of renewal phenomena. Furthermore, it can be applied in cases where the cluster structure is an important factor in the analysis.

The proposed methods perform well in real spatial contexts, showing appealing features which can be useful to practitioners in important scientific fields such as contagion analysis and general spatiotemporal analysis. Other choices of random measures potentially lead to similar outcomes. The study of these more general classes of priors will be pursued elsewhere.

Acknowledgements We thank the editor and two anonymous reviewers for their useful comments which significantly improved the presentation and quality of the paper. The first author is grateful to Prof. Ramsés Mena for the valuable suggestions on an earlier version of the manuscript. This research was partially supported by a DGAPA Postdoctoral Scholarship.

References

1. Aldous, D.J.: Exchangeability and related topics. In: École d'Été de Probabilités de Saint-Flour XIII 1983, pp. 1–198. Springer, Berlin (1985). https://doi.org/10.1007/BFb0099421
2. Andersson, H., Britton, T.: Stochastic Epidemic Models and Their Statistical Analysis. Lecture Notes in Statistics, vol. 151. Springer, New York, NY (2000)
3. Blackwell, D., MacQueen, J.B.: Ferguson distributions via polya urn schemes. Ann. Stat. (1973). https://doi.org/10.1214/aos/1176342372
4. Bladt, M., Nielsen, B.F.: Matrix-Exponential Distributions in Applied Probability. Probability Theory and Stochastic Modelling, vol. 81. Springer, Boston (2017). https://doi.org/10.1007/978-1-4939-7049-0
5. Brauer, F., van den Driessche, P., Wu, J. (eds.): Mathematical Epidemiology. Lecture Notes in Mathematics, vol. 1945. Springer, Heidelberg (2008). https://doi.org/10.1007/978-3-540-78911-6
6. Bulla, P., Muliere, P.: Bayesian nonparametric estimation for reinforced markov renewal processes. Stat. Inference Stoch. Process. **10**(3), 283–303 (2007). https://doi.org/10.1007/s11203-006-9000-x
7. Coen, A., Gutierrez, L., Mena, R.H.: Modeling failures times with dependent renewal type models via exchangeability. Submited (2018)
8. Coen, A., Mena, R.H.: Ruin probabilities for Bayesian exchangeable claims processes. J. Stat. Plan. Inference **166**, 102–115 (2015). https://doi.org/10.1016/J.JSPI.2015.01.005
9. Cori, A., Nouvellet, P., Garske, T., Bourhy, H., Nakouné, E., Jombart, T.: A graph-based evidence synthesis approach to detecting outbreak clusters: an application to dog rabies. PLOS Comput. Biol. **14**(12), e1006554 (2018). https://doi.org/10.1371/journal.pcbi.1006554
10. Doksum, K.: Tailfree and neutral random probabilities and their posterior distributions. Ann. Probab. **2**(2), 183–201 (1974)
11. Escobar, M.D.: Estimating normal means with a Dirichlet process prior. J. Am. Stat. Assoc. (1994). https://doi.org/10.1080/01621459.1994.10476468
12. Escobar, M.D., West, M.: Bayesian density estimation and inference using mixtures. J. Am. Stat. Assoc. **90**(430), 577 (1995). https://doi.org/10.2307/2291069
13. Escobar, M.D., West, M.: Bayesian density estimation and inference using mixtures. J. Am. Stat. Assoc. (1995). https://doi.org/10.1080/01621459.1995.10476550
14. Ewens, W.J.: The sampling theory of selectively neutral alleles. Theor. Popul. Biol. **3**(1), 87–112 (1972). https://doi.org/10.1016/0040-5809(72)90035-4
15. Ferguson, T.S.: A bayesian analysis of some nonparametric problems. Ann. Stat. **1**(2), 209–230 (1973). https://doi.org/10.1214/aos/1176342360

16. Ferguson, T.S.: Bayesian density estimation by mixtures of normal distributions. Recent Advances in Statistics, pp. 287–302 (1983). https://doi.org/10.1016/B978-0-12-589320-6.50018-6
17. Frees, E.W.: Nonparametric renewal function estimation. Ann. Stat. **14**(4), 1366–1378 (1986). https://doi.org/10.1214/aos/1176350163
18. Gebizlioglu, O.L., Eryilmaz, S.: The maximum surplus in a finite-time interval for a discrete-time risk model with exchangeable, dependent claim occurrences. Appl. Stoch. Model. Bus. Ind. (2018). https://doi.org/10.1002/asmb.2415
19. Ghosal, S., van der Vaart, A.: Posterior convergence rates of Dirichlet mixtures at smooth densities. Ann. Stat. **35**(2), 697–723 (2007). https://doi.org/10.1214/009053606000001271
20. Görür, D., Rasmussen, C.E.: Dirichlet process gaussian mixture models: choice of the base distribution. J. Comput. Sci. Technol. (2010). https://doi.org/10.1007/s11390-010-9355-8
21. Ishwaran, H., James, L.F.: Gibbs sampling methods for stick-breaking priors. J. Am. Stat. Assoc. (2001). https://doi.org/10.1198/016214501750332758
22. Kingman, J.F.C.: Poisson Processes. Oxford Studies in Probability. Clarendon Press, Oxford (1992)
23. Korwar, R.M., Hollander, M.: Contributions to the theory of dirichlet processes. Ann. Probab. **1**(4), 705–711 (1973). https://doi.org/10.1214/aop/1176996898
24. Kovalenko, I.N., Pegg, P.A.: Mathematical Theory of Reliability of Time Dependent Systems With Practical Applications. Wiley Series in Probability and Statistics: Applied Probability and Statistics. Wiley, Hoboken (1997)
25. Lenk, P.J.: The logistic normal distribution for bayesian, nonparametric, predictive densities. J. Am. Stat. Assoc. **83**(402), 509 (1988). https://doi.org/10.2307/2288870
26. Lijoi, A., Mena, R.H., Prünster, I.: Bayesian nonparametric analysis for a generalized dirichlet process prior. Stat. Inference Stoch. Process. **8**(3), 283–309 (2005). https://doi.org/10.1007/s11203-005-6071-z
27. Lo, A.Y.: On a class of bayesian nonparametric estimates: I. density estimates. Ann. Stat. **12**(1), 351–357 (1984). https://doi.org/10.1214/aos/1176346412
28. Maceachern, S.N., Müller, P.: Estimating mixture of dirichlet process models. J. Comput. Graph. Stat. (1998). https://doi.org/10.1080/10618600.1998.10474772
29. Nadarajah, S., Li, R.: The exact density of the sum of independent skew normal random variables. J. Comput. Appl. Math. **311**, 1–10 (2017). https://doi.org/10.1016/J.CAM.2016.06.032
30. Nadarajaha, S., Chanb, S.: The exact distribution of the sum of stable random variables. J. Comput. Appl. Math. **349**, 187–196 (2019). https://doi.org/10.1016/J.CAM.2018.09.044
31. Neal, R.M.: Markov chain sampling methods for dirichlet process mixture models. J. Comput. Graph. Stat. (2000). https://doi.org/10.1080/10618600.2000.10474879
32. Pitman, J.: Some developments of the Blackwell-MacQueen urn scheme. In: Statistics, Probability and Game Theory, pp. 245–267. Institute of Mathematical Statistics, Beachwood (1996). https://doi.org/10.1214/lnms/1215453576
33. Pitman, J.: Poisson-Kingman partitions. In: Statistics and Science: A Festschrift for Terry Speed, pp. 1–34. Institute of Mathematical Statistics, Beachwood (2003). https://doi.org/10.1214/lnms/1215091133
34. Prünster, I., Lijoi, A., Regazzini, E.: Distributional results for means of normalized random measures with independent increments. Ann. Stat. **31**(2), 560–585 (2003). https://doi.org/10.1214/aos/1051027881
35. Rolski, T., Schmidli, H., Schmidt, V., Teugels, J.: Stochastic processes for insurance and finance. Wiley Series in Probability and Statistics. Wiley, Chichester (1999). https://doi.org/10.1002/9780470317044
36. Tokdar, S.T.: Posterior consistency of Dirichlet location-scale mixture of normals in density estimation and regression. Sankhy Indian J. Stat. (2003–2007) **68**(1), 90–110 (2006)
37. Webber, R.: Communicable Disease Epidemiology and Control: A Global Perspective. Modular Texts, Cabi (2009)

38. World Health Organization: WHO Expert Consultation on Rabies: Third Report. World Health Organization, Geneva (2018). 92 4 120931 3
39. Xiao, S., Kottas, A., Sansó, B.: Modeling for seasonal marked point processes: an analysis of evolving hurricane occurrences. Ann. Appl. Stat. **9**(1), 353–382 (2015). https://doi.org/10.1214/14-AOAS796
40. Ypma, R.J.F., Donker, T., van Ballegooijen, W.M., Wallinga, J.: Finding evidence for local transmission of contagious disease in molecular epidemiological datasets. PLoS ONE **8**(7), e69875 (2013). https://doi.org/10.1371/journal.pone.0069875
41. Zhang, C.H.: Estimation of sums of random variables: examples and information bounds. Ann. Stat. **33**(5), 2022–2041 (2005). https://doi.org/10.1214/009053605000000390

An Efficient Method to Determine the Degree of Overlap of Two Multivariate Distributions

Eduardo Gutiérrez-Peña and Stephen G. Walker

Abstract Assessing the degree to which two probability density functions overlap is an important problem in several applications. Most of the existing proposals to tackle this problem can only deal with univariate distributions. For multivariate problems, existing methods often rely on unrealistic parametric distributional assumptions or are such that the corresponding univariate marginal measures are combined using ad hoc procedures. In this paper, we propose a new empirical measure of the degree of overlap of two multivariate distributions. Our proposal makes no assumptions on the form of the densities and can be efficiently computed even in relatively high-dimensional problems.

Keywords Distance matrix · Crossmatch algorithm · Multivariate analysis

1 Introduction and Motivation

Assessing the degree to which two probability density functions overlap is an important problem in several applications. For instance, in ecology this problem appears when comparing ecological niches (see, for example, Ridout and Linkie [8], Geange et al. [2], Swanson et al. [10]). In medical settings, it is of interest to study the performance of diagnostic tests, where it is necessary to compare the distribution of the scores of healthy patients with that corresponding to sick patients [6]. In the context of causal inference, imbalance occurs if the distributions of relevant pretreatment variables differ for the treatment and control groups; lack of complete overlap occurs if there are regions in the space of relevant pretreatment variables where there are treated units but no controls, or controls but no treated units. These are both important

E. Gutiérrez-Peña (✉)
National Autonomous University of Mexico, Mexico City, Mexico
e-mail: eduardo@sigma.iimas.unam.mx

S. G. Walker
University of Texas at Austin, Austin, USA
e-mail: s.g.walker@math.utexas.edu

I. Antoniano-Villalobos et al. (eds.), *Selected Contributions on Statistics and Data Science in Latin America*, Springer Proceedings in Mathematics & Statistics 301,
https://doi.org/10.1007/978-3-030-31551-1_5

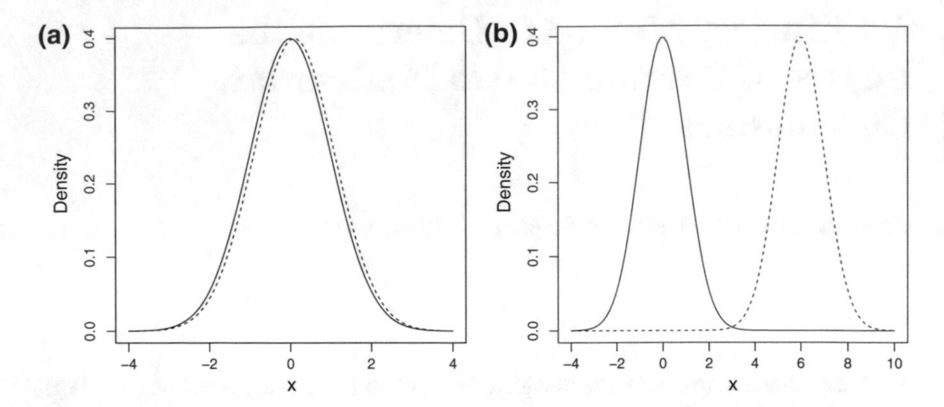

Fig. 1 An example of two probability densities showing: **a** complete overlap; **b** no overlap

issues because they force the analyst to rely more heavily on model specification and less on direct support from the data (Gelman and Hill [3], Chap. 10).

The idea behind most of the measures of overlap starts by assuming that a set of d-dimensional observations $\boldsymbol{x}_{1,1}, \ldots, \boldsymbol{x}_{1,n_1}$ is available. If these data can be regarded as a sample of a continuous random vector $\boldsymbol{X}$, then we can use them to estimate the probability density function $f_1(\boldsymbol{x}_1)$. Similarly, if $\boldsymbol{x}_{2,1}, \ldots, \boldsymbol{x}_{2,n_2}$ are d-dimensional observations on the use of the same resources by another species, we can proceed in the same way and estimate the corresponding density function $f_2(\boldsymbol{x}_2)$. It is illustrative to consider the following two cases. If f_1 and f_2 are equal, then they would fully overlap. Conversely, if these density functions are completely different—in particular, if the intersection of their supports is empty—then there would be no overlap at all. Figure 1 illustrates these two situations. The problem of quantifying an overlap thus reduces to the choice of a measure of similarity between two density functions. In the case of discrete random variables, we would compare the corresponding probability mass functions instead.

In the univariate case ($d = 1$), a number of measures of overlap have been proposed based on this approach. For example, Ridout and Linkie [8] mention the following three measures that can be used to quantify the overlap of the two densities f_1 and f_2 in ecological settings:

$$\Delta(f_1, f_2) = \int \min\{f_1(x), f_2(x)\}\, \mathrm{d}x,$$

$$\rho(f_1, f_2) = \int \sqrt{f_1(x) f_2(x)}\, \mathrm{d}x,$$

and

$$\lambda(f_1, f_2) = 2 \int f_1(x) f_2(x)\, \mathrm{d}x \Bigg/ \left\{ \int f_1(x)^2\, \mathrm{d}x + \int f_2(x)^2\, \mathrm{d}x \right\}.$$

Perhaps the most popular of these measures is the *coefficient of overlapping*, $\Delta(f_1, f_2)$, which is related to the total variation distance. (Note that $\rho(f_1, f_2)$ is sim-

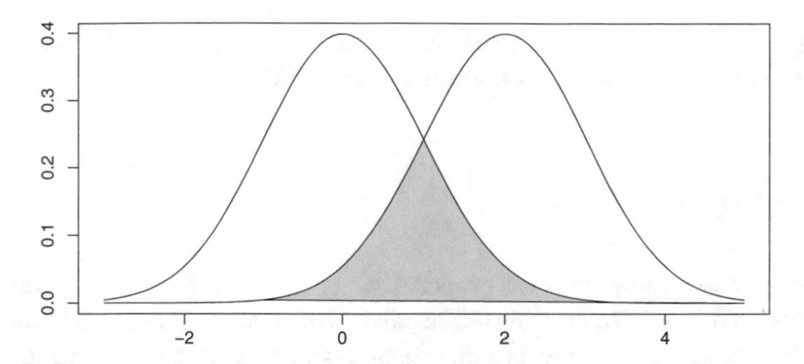

Fig. 2 Graphical depiction of the coefficient of overlapping

ilarly related to the Hellinger distance.) For one thing, the interpretation of $\Delta(f_1, f_2)$ as the area under the minimum of f_1 and f_2 is intuitively appealing (see Fig. 2). This coefficient is widely used by ecologists to describe daily animal activity patterns thanks to the availability of user-friendly software such as the R package overlap [5]; see Sect. 4. For these reasons, in this paper, we will only compare our measure of overlap with $\Delta(f_1, f_2)$.

Once the density estimates $\hat{f}_1$ and $\hat{f}_2$ are obtained, we can estimate the coefficient of overlapping by $\Delta(\hat{f}_1, \hat{f}_2)$; this estimate will be denoted by CO. Unfortunately, the CO is difficult to compute in multivariate settings. To get around this problem, Geange et al. [2] proposed to estimate the coefficient of overlapping separately for each of the variables and then take a simple average of the resulting measures. However, this procedure does not take into account the correlation between the variables and is not invariant under certain transformations (such as rotations) that preserve the degree of overlap. Swanson et al. [10] propose a new probabilistic method for determining the niche region and pairwise niche overlap that can be extended beyond two dimensions and produces consistent and unique bivariate projections of multivariate data. However, their method relies on the assumption of normality.

In this paper, we bring together some known elements of testing for the equality between two multivariate distributions [9] and propose to use the output of the test to measure the degree of overlap. This overlap measure has been studied as a consequence of looking into the asymptotic properties of the crossmatch test. An unused by-product of implementing the test is that the crossmatched pairs are actually provided in the construction, which estimates the measure of overlap between the two distributions.

Our proposal makes no assumptions on the form of the densities and can be efficiently computed even in relatively high-dimensional problems. Moreover, it can also deal with circular and directional data.

In the next section, we describe our proposal and compare it with the coefficient of overlapping in the case of univariate normal distributions. In Sect. 3, we present a small simulation study to illustrate how our measure of overlap behaves in a range of bivariate settings where the distributions can be asymmetric and/or multimodal. Finally, in Sect. 4, we provide some concluding remarks.

2 The Degree of Overlap

The measure of overlap that we propose, which we term *Degree of Overlap* and denote by *DO*, is defined in terms of the pairwise distances between all of the points in the two samples. While in principle any distance could be used, in this paper we will only consider the Euclidean distance, and if the variables are measured in different units, each of them can be standardized to get unitless measurements before computing the distances.

Consider the two data sets, $\boldsymbol{x}_{1,1}, \ldots, \boldsymbol{x}_{1,n_1}$ and $\boldsymbol{x}_{2,1}, \ldots, \boldsymbol{x}_{2,n_2}$, stacked one below the other and arranged in a $(n_1 + n_2) \times d$ matrix in such a way that the first n_1 rows correspond to the observations from f_1, while the last n_2 rows correspond to the observations from f_2. Now compute all the pairwise distances between the $(n_1 + n_2)$ data points (rows) and arrange them as usual in an $(n_1 + n_2) \times (n_1 + n_2)$ *distance matrix* D with zeros along the diagonal. Note that D can be written in block form as

$$D = \begin{pmatrix} D_{11} & D_{12} \\ D_{21} & D_{22} \end{pmatrix},$$

where D_{11} is an $n_1 \times n_1$ matrix containing the pairwise distances between the observations from f_1, D_{22} is an $n_2 \times n_2$ matrix containing the distances between the observations from f_2, and D_{12} is an $n_1 \times n_2$ matrix containing the cross-distances between observations from f_1 and f_2. If f_1 and f_2 do not overlap at all, we can expect the distances in the matrices D_{11} and D_{22} to be smaller than those in the matrix D_{12} (equivalently, D_{21}, due to the symmetry of D). Conversely, if f_1 and f_2 overlap completely, the distances in D_{11} and D_{22} would be of the same order as those in D_{12} (or D_{21}).

Let $\nu : \{1, 2, \ldots, n_1 + n_2\} \rightarrow \{1, 2, \ldots, n_1 + n_2\}$ be the mapping that matches each observation uniquely with another observation by minimizing the sum of the distances between pairs, subject to the conditions $\nu(i) \neq i$ and $\nu(\nu(i)) = i$. Now, for each $i = 1, 2, \ldots, n_1$, let $\delta_1(i) = 1$ if $\nu(i) \leq n_1$ and $\delta_1(i) = 0$ otherwise; and for $i = n_1 + 1, \ldots, n_1 + n_2$, let $\delta_2(i) = 1$ if $\nu(i) > n_1$ and $\delta_2(i) = 0$ otherwise. Finally, let $m_1 = \sum_{i=1}^{n_1} \delta_1(i)$ and $m_2 = \sum_{i=n_1+1}^{n_1+n_2} \delta_2(i)$. Focusing for the moment on the observations from f_1, if f_1 and f_2 do not overlap then we can expect m_1 to be approximately n_1, i.e., $p = 100\%$ of the observations from f_1 would be matched with observations from the same distribution f_1. However, if f_1 and f_2 overlap completely we would only expect roughly $p = n_1/(n_1 + n_2) \times 100\%$ of the observations from f_1 to be matched with observations from the same distribution f_1. In this case, m_1 would be approximately $n_1^2/(n_1 + n_2)$. A similar argument can of course be made for the observations from f_2.

As is the case with most of the measures of overlap proposed in the literature, we would like our degree of overlap to take the value 0 for no overlap and the value 1 for a complete overlap. We define the *Degree of Overlap* of f_1 and f_2 as

$$DO = \min\left\{1, \frac{n\,(n_1 - m_1)}{n_1\, n_2}\right\},$$

where $n = n_1 + n_2$ (note that $n_1 - m_1 = n_2 - m_2$). Clearly, $0 \leq DO \leq 1$. Asymptotically, the upper limit for $n(n_1 - m_1)/(n_1 n_2)$ is 1; hence, our inclusion of the minimum with 1. Values of $n(n_1 - m_1)/(n_1 n_2) > 1$ are very unlikely; they would indicate a sampling anomaly suggestive of perfect overlap.

This measure was inspired by the crossmatch test statistic (i.e., $(n_1 - m_1)/n$), proposed by Rosenbaum [9] to test for the equality of two multivariate distributions. It is related to the following quantity:

$$\int \frac{\pi(1-\pi) f_1(x) f_2(x)}{\pi f_1(x) + (1-\pi) f_2(x)}\, \mathrm{d}x,$$

where it is assumed that $n_1/(n_1 + n_2) \to \pi$ for some $\pi \in (0, 1)$; that is, the sample sizes are assumed to be comparable; see Arias-Castro and Pelletier [1].

In the remainder of this section, we compare our *DO* with $\Delta(f_1, f_2)$, the coefficient of overlapping defined in the previous section. Here we will only be dealing with two univariate normal distributions, and so it is easy to compute the true value of the coefficient of overlapping. However, for the sake of simplicity, in what follows we will still denote this measure by *CO* even though we are no longer referring to an estimate. This explains why, in Figs. 3 and 4, the curves for *CO* are smooth while those for *DO* are not: the values of *DO* were estimated based on samples of sizes $n_1 = n_2 = 1000$.

For the following examples, as well as for those in the next section, we used the statistical environment R [7]. Specifically, in order to compute number of crossmatches $n_1 - m_1$, we used the function `crossmatchtest` of the R package crossmatch [4].

We first compare the *DO* with the *CO* in the context of two univariate normal distributions with the same variance but with varying levels of overlaps given by shifts in the mean. Specifically, Fig. 3 shows the results of the comparison as $f_2 = N(s, 1)$ is shifted to the right away from $f_1 = N(0, 1)$ by a given number s of standard deviations. Similarly, Fig. 4 shows the results of the comparison as the dispersion of $f_2 = N(0, s^2)$ is increased relative to that of $f_1 = N(0, 1)$ in terms of a given number s of standard deviations.

In both of these cases, we can see that the two measures are strongly positively correlated. As expected, both *DO* and *CO* decrease as the level of overlap decreases. It is worth stressing that, in practice, the *DO* can be just as easily computed even for relatively high-dimensional, while computing the *CO* would require the estimation of two multivariate densities as well as computing the area under the minimum of such estimates.

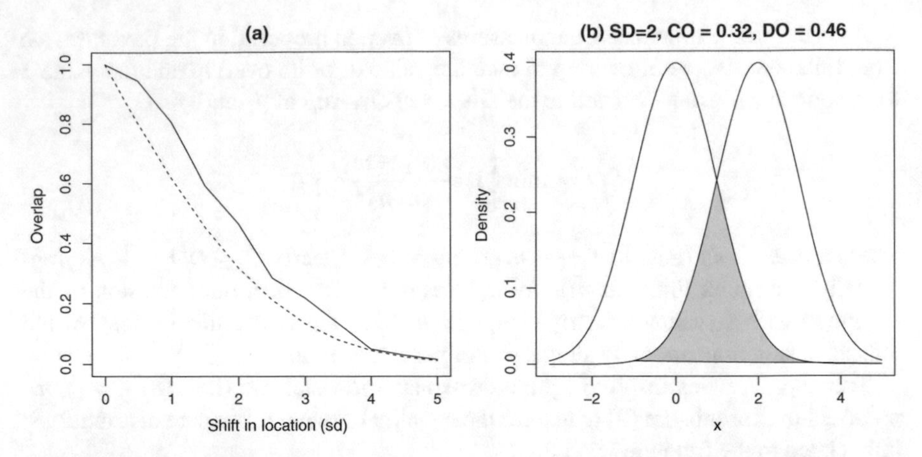

Fig. 3 Comparison of the degree of overlap (DO) and the coefficient of overlapping (CO) as the location of f_2 is shifted from f_1 by a given number of standard deviations: **a** DO (continuous line) and CO (dashed line) as a function of the shift; **b** illustrative example: overlap between $f_1 = N(0, 1)$ and $f_2 = N(2, 1)$

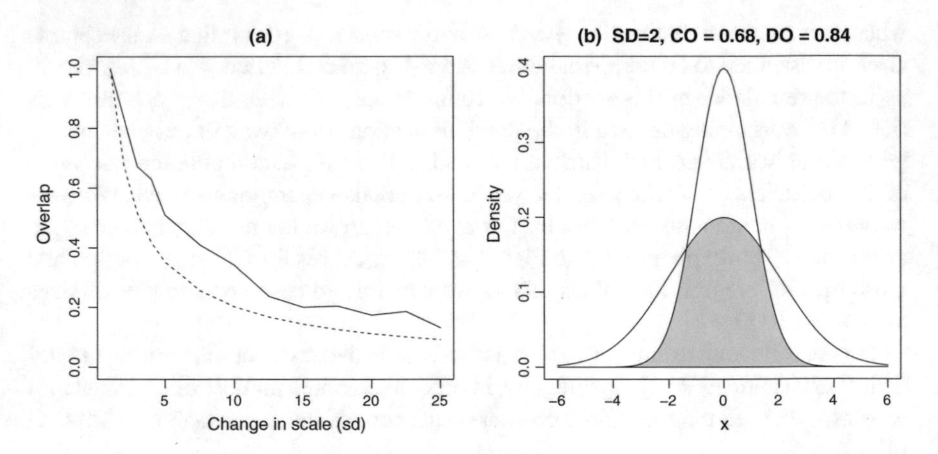

Fig. 4 Comparison of the degree of overlap (DO) and the coefficient of overlapping (CO) as the standard deviation of f_2 increases relative to that of f_1: **a** DO (continuous line) and CO (dashed line) as a function of the standard deviation; **b** illustrative example: overlap between $f_1 = N(0, 1)$ and $f_2 = N(0, 4)$

3 Simulation Study

In this section, we will study the performance of the *DO* measure in a range of bivariate cases. Note that, even in these simple examples, computing the *CO* would be far from easy. As an alternative, and as pointed out in Sect. 1, a simple proposal for such situations is to estimate univariate measures (such as the *CO*, the *DO*, or either of the other measures described in the previous section) for each of the variables

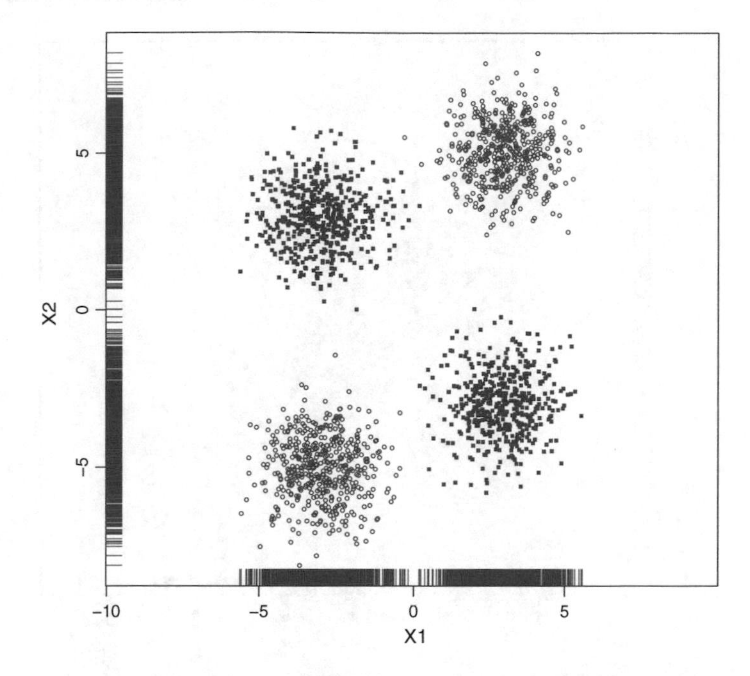

Fig. 5 Two mixtures of bivariate normal distributions. The bivariate overlap is very low ($DO = 0.002$). On the other hand, the overlap along the X_1 axis is very high ($DOX1 = 1.0$), while that along the X_2 axis can be regarded as moderate ($DOX2 = 0.4$)

and then take a simple average of the resulting values. This procedure has several limitations, as will be shown in the following examples by comparing the actual bivariate *DO* with the average of the corresponding univariate measures.

3.1 Nonoverlapping Mixtures of Bivariate Normal Distributions

We simulated a sample of size $n_1 = 1000$ and sample of size $n_2 = 1000$ from two different mixtures of bivariate normal distributions. The data are shown in Fig. 5. In this case, the bivariate degree of overlap is $DO(X_1, X_2) = 0.002$. On the other hand, the degree of overlap along the X_1 axis is $DO(X_1) = 1.0$, while that along the X_2 axis is $DO(X_2) = 0.4$. These values are consistent with what can be observed in Fig. 5; namely, the bivariate degree of overlap is very low, while the univariate degrees of overlap for X_1 and X_2 are quite high and relatively small, respectively.

Note that, in this case, the simple procedure proposed by Geange et al. [2] would produce a relatively large value for the “bivariate overlap” (0.7), since it would take a simple average of the two univariate measures. Note also that rotating the two data sets by the same angle should not affect the degree of overlap. Our measure of overlap

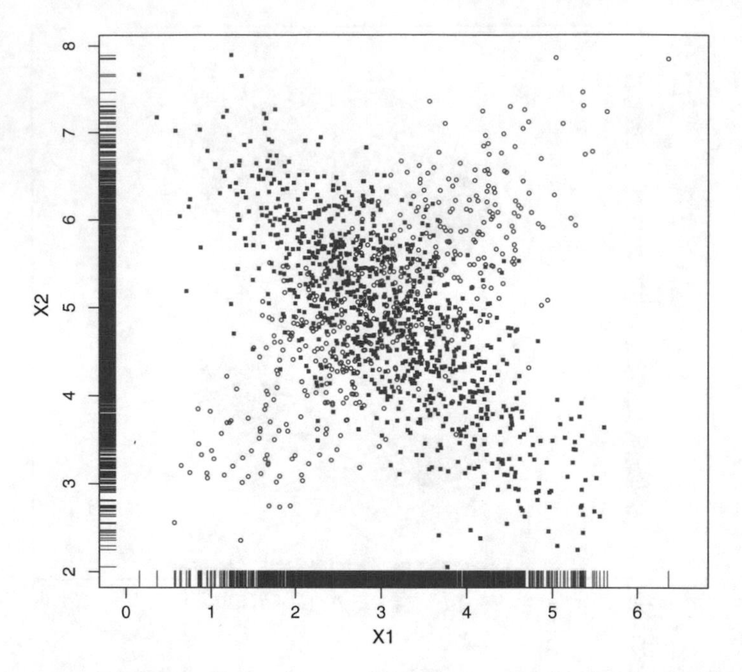

Fig. 6 Two bivariate normal distributions with different correlation coefficients. The bivariate overlap takes a moderate value ($DO = 0.55$). However, the overlap along each of the X_1 and X_2 axes is quite high ($DOX1 = 1.0$, $DOX2 = 0.96$)

remains the same because the distances between the data points do not change with any rotation. However, the method proposed by Geange et al. [2] would produce different values for each rotation because it depends on the corresponding marginal measures, and both marginal distributions (projections) may change dramatically when the data are rotated.

3.2 Overlapping Bivariate Normal Distributions

For this example, we simulated a sample of size $n_1 = 500$ from a bivariate normal distribution with a positive correlation coefficient, and sample of size $n_2 = 1000$ from a bivariate normal distribution with a negative correlation coefficient. The data are shown in Fig. 6. In this case the bivariate degree of overlap is $DO(X_1, X_2) = 0.55$. However, the degree of overlap along each of the two axes is quite high, with $DO(X_1) = 1.0$ and $DO(X_2) = 0.96$. As in the previous example, these values are consistent with what we can observe in the figure. Note also that the method of Geange et al. [2] would again provide misleading results in this case. Specifically, their method would suggest a high degree of "bivariate overlap" (0.98).

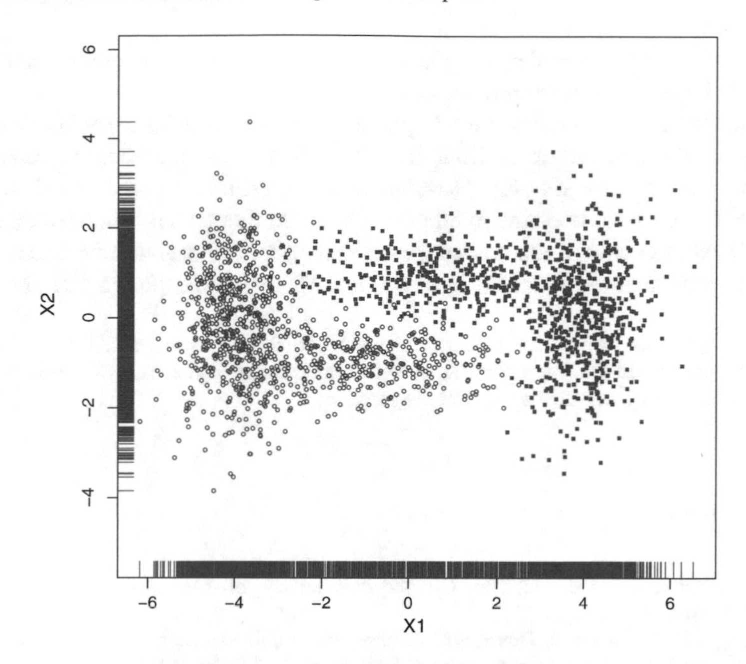

Fig. 7 Two mixtures of bivariate normal distributions yielding unimodal but asymmetric distributions. The bivariate overlap is very low ($DO = 0.058$). On the other hand, the overlap along the X_1 axis is somewhat low ($DOX1 = 0.32$), while that along the X_2 axis is relatively high ($DOX2 = 0.81$)

3.3 Asymmetric Bivariate Distributions

For this last example, we simulated a sample of size $n_1 = 1000$ and sample of size $n_2 = 1000$ from two different mixtures of bivariate normal distributions, with parameters suitably chosen so as to produce unimodal but asymmetric distributions. The data are shown in Fig. 7. In this case, the bivariate degree of overlap is $DO(X_1, X_2) = 0.058$. On the other hand, the degree of overlap along the X_1 axis is $DO(X_1) = 0.32$, while that along the X_2 axis is $DO(X_2) = 0.81$.

As in the previous examples, the simple procedure proposed by Geange et al. [2] would produce a value for the "bivariate overlap" (0.565) that does not really reflect the actual degree of overlap.

4 Concluding Remarks

In this paper, we have proposed a novel distance-based method to quantify the degree of overlap of two multivariate distributions. Our proposal makes no assumptions about the form of the two density functions and can be efficiently computed even in relatively high-dimensional problems.

Further work is required in order to provide full inferences for the actual degree of overlap of two multivariate distributions. The *DO* measure introduced in this paper provides a point estimate only. However, interval estimates could in principle be obtained by means of bootstrap methods. Also, even though the simulated examples above illustrate some of the advantages of our proposal, it would be necessary to study its performance in real-life applications. This will be explored elsewhere.

Acknowledgements The work of the first author was partially supported by the *Sistema Nacional de Investigadores*, Mexico. The authors are grateful for the comments and suggestions of the editor and two reviewers on an earlier version of the paper.

References

1. Arias-Castro, E., Pelletier, B.: On the consistency of the crossmatch test (2015). arXiv:1509.05790 [math.ST]
2. Geange, S.W., Pledger, S., Burns, K.C., Shima, J.S.: A unified analysis of niche overlap incorporating data of different types. Methods Ecol. Evol. **2**, 175–184 (2011)
3. Gelman, A., Hill, J.: Data Analysis Using Regression and Multilevel/Hierarchical Models. Cambridge University Press, Cambridge (2007)
4. Heller, R., Small, D., Rosenbaum, P.: `crossmatch`: the cross-match test. R package version 1.3-1 (2012). https://CRAN.R-project.org/package=crossmatch
5. Meredith, M., Ridout, M.: Overlap: estimates of coefficient of overlapping for animal activity patterns. R package version 0.3.2 (2018). https://CRAN.R-project.org/package=overlap
6. Pepe, M.S.: The Statistical Evaluation of Medical Tests for Classification and Prediction. University Press, Oxford (2003)
7. R Core Team: R: A Language and Environment for Statistical Computing. R Foundation for Statistical Computing, Vienna, Austria (2018). https://www.R-project.org/
8. Ridout, M.S., Linkie, M.: Estimating overlap of daily activity patterns from camera trap data. J. Agric. Biol. Environ. Stat. **14**, 322–337 (2009)
9. Rosenbaum, P.R.: An exact distribution-free test comparing two multivariate distributions based on adjacency. J. R. Stat. Soc. B **67**, 515–530 (2005)
10. Swanson, H.K., Lysy, M., Power, M., Stasko, A.D., Johnson, J.D., Reist, J.D.: A new probabilistic method for quantifying n-dimensional ecological niches and niche overlap. Ecology **96**, 318–324 (2015)

Clustering via Nonsymmetric Partition Distributions

Asael Fabian Martínez

Abstract Random partition models are widely used to perform clustering, since their features make them appealing options. However, additional information regarding group properties is not straightforward to incorporate under this approach. In order to overcome this difficulty, a novel approach to infer about clustering is presented. By relaxing the symmetry property of random partitions' distributions, we are able to include group sizes in the computation of the probabilities. A Bayesian model is also given, together with a sampling scheme, and it is tested using simulated and real datasets.

Keywords Bayesian modeling · Density estimation · Ordered set partitions

1 Introduction

The problem of clustering has received great attention in different fields of research, among other facts, due to its generality and, as a consequence, its applicability. The elemental idea behind clustering is to gather a given set of items in such a way that those belonging to one group are similar among them and, at the same time, are dissimilar to items in other groups. The degree of (dis)similarity can be defined by some metric or according to some probabilistic model assigned to each group.

When formulating clustering as an inference problem, the parameter space poses some challenges, for example, in the choice of nontrivial *prior* distributions under a Bayesian perspective. The theory of random partitions is a widely used approach adapted to perform Bayesian analysis; however, existing distributions over partitions do not give further information on other group properties.

In this paper, we present and explore a novel approach to infer about clustering based on random partitions, but where it is possible to obtain group structures sat-

A. F. Martínez (✉)
Departamento de Matemáticas, Universidad Autónoma Metropolitana, Unidad Iztapalapa, Av. San Rafael Atlixco 186, 09340 Mexico City, Mexico
e-mail: fabian@xanum.uam.mx

I. Antoniano-Villalobos et al. (eds.), *Selected Contributions on Statistics and Data Science in Latin America*, Springer Proceedings in Mathematics & Statistics 301,
https://doi.org/10.1007/978-3-030-31551-1_6

isfying an ordering constraint. This is achieved by relaxing the symmetry property of exchangeable random partitions. In Sect. 2, the proposed model is derived, and in Sect. 3, a nonsymmetric partition distribution is presented. A characteristic of this distribution is that it takes into account the size of groups in the computation of probabilities, favoring partitions whose groups are ordered by size.

Once the model is fully developed, in Sect. 4, conditional distributions are derived and a sampling scheme is provided. It is also explained how to compute the posterior distribution for the number of groups, something feasible since our approach is partition based, and how to estimate a mixture density. Afterward, in Sect. 5, two datasets are used to test the performance of the model. Finally, some final remarks are given in Sect. 6.

2 A Bayesian Model for Clustering

Suppose there is a set of observations $y = (y_1, \ldots, y_n)$ we wish to cluster. Any possible arrangement of y, or clustering structure, can be represented by a partition $G = \{G_1, \ldots, G_k\}$ of y, for some positive k. Under a model-based clustering approach, all observations in G_j, for a fixed j, follow the same probabilistic model, say $\kappa(x_j)$.

A common representation for clustering has been exploited by random partition clustering models found, for example, in Bayesian nonparametric models (see, for example, [2] for further references). In this case, the underlying random partition is defined in the set of partitions, $\mathscr{P}_{[n]}$, and its distribution usually belongs to the class of *exchangeable partition probability functions* (EPPFs).

Every element $\pi \in \mathscr{P}_{[n]}$ is a partition of the set $[n] := \{1, 2, \ldots, n\}$ into k non-empty subsets, also called *blocks*, π_j, for some k. From this, every clustering structure G is encoded by observations indices, for example, clustering $\{\{y_1\}, \{y_2, y_3\}\}$ corresponds to the partition $\{1\}/\{2, 3\}$.

On the other hand, an EPPF g has two properties: first, the probability of any partition is function of its blocks sizes and second, g is a symmetric function. Consider the case $n = 3$, where $\#\mathscr{P}_{[3]} = 5$. First property implies that

$$\Pr(\Pi = \{1, 2\}/\{3\}) = g(2, 1) = \Pr(\Pi = \{1, 3\}/\{2\}),$$

whereas the second one makes additionally

$$\Pr(\{1, 2\}/\{3\}) = g(2, 1) = g(1, 2) = \Pr(\{1\}/\{2, 3\}).$$

It is important to highlight that blocks in π are indistinguishable for any $\pi \in \mathscr{P}_{[n]}$. Thus, for example, partition $\{2, 3\}/\{1\}$ is equivalent to $\{1\}/\{2, 3\}$, and furthermore, even if they could be treated as different elements, the symmetry of g would assign the same probability.

A second approach for clustering, naturally derived from mixture models, makes use of an indicator vector $d = (d_1, \ldots, d_n)$ in such a way that each group G'_j is defined as $G'_j = \{y_i : d_i = j\}$, for some k', and in such a case, $y_i \sim \kappa(x_{d_i})$, independently, for $i = 1, \ldots, n$. Despite the fact that clustering structures $G' = \{G'_1, \ldots, G'_{k'}\}$ and G, defined above, for a common dataset y, are frequently considered as the same, there are some differences. On the one hand, it is possible to have empty sets in G', and on the other hand, clustering structures G' like $\{\{y_2, y_3\}, \{y_1\}\}$ and $\{\{y_1\}, \{y_2, y_3\}\}$ have a different meaning. Indeed, the underlying support when using the indicator vector d is known as the combinatorial class of *words*, $\mathscr{W}_{[n]}$. In [3], a more complete and formal study of these two spaces is provided.

Although there are also different proposals defined over the space $\mathscr{W}_{[n]}$ to perform clustering (see, for example, [5, 6]), they allow to have empty groups which forces the use of some additional procedure to remove them, and limit its applicability in nonparametric settings where there is a potentially infinite number of groups.

Based on the two spaces and their modeling abilities, we will present a clustering model defined over a different space, the so-called *ordered set partitions*. This combinatorial class, denoted hereafter as $\mathscr{OP}_{[n]}$, is obtained from the class $\mathscr{W}_{[n]}$ after removing empty groups. Equivalently, it can be obtained from $\mathscr{P}_{[n]}$ when every permutation of the blocks in any of its elements is considered as a different partition, i.e., if $\pi \in \mathscr{P}_{[n]}$, with $\pi = \pi_1/\cdots/\pi_k$, then $\pi_{\rho(1)}/\cdots/\pi_{\rho(k)}$ belongs to $\mathscr{OP}_{[n]}$ for every permutation ρ of $[k]$. A further explanation of these derivations and the properties of $\mathscr{OP}_{[n]}$ are also studied in [3].

Therefore, our clustering model is defined as follows. Let μ_0 be a probability distribution over $\mathscr{OP}_{[n]}$, Π be a μ_0-distributed random partition, κ be a density function, with parameter x_j, modeling all observations belonging to group j, and ν_0 be the prior distribution for the kernel parameter. The model can be written hierarchically as

$$\begin{aligned}
y_i|\Pi = \pi, X = x &\overset{\text{ind.}}{\sim} \kappa(x_j)\mathbf{1}(i \in \pi_j), && i = 1, \ldots, n \qquad (1)\\
X_j|\Pi = \pi &\overset{\text{i.i.d.}}{\sim} \nu_0(\phi), && j = 1, \ldots, \#\pi\\
\Pi &\sim \mu_0(\psi),
\end{aligned}$$

where $\pi = \pi_1/\ldots/\pi_k \in \mathscr{OP}_{[n]}$ for some $1 \leq k \leq n$, $\#\pi$ denotes the number of groups in the partition π, and ϕ and ψ are finite dimensional parameters. The likelihood function is given by

$$p(y|\pi, x) = \prod_{j=1}^{\#\pi}\prod_{i=1}^{n} \kappa(y_i; x_j)\mathbf{1}(i \in \pi_j) = \prod_{j=1}^{\#\pi}\prod_{i\in\pi_j} \kappa(y_i; x_j),$$

and the joint posterior distribution for (π, x) is

$$p(\pi, x|y) \propto \prod_{j=1}^{\#\pi}\prod_{i\in\pi_j} \kappa(y_i; x_j)\nu_0(x_j; \phi)\ \mu_0(\pi; \psi) = \prod_{j=1}^{\#\pi} \ell(y; \pi_j, x_j, \phi)\ \mu_0(\pi; \psi), \qquad (2)$$

where ℓ is the likelihood of cluster π_j. If there is interest only in the clustering structure π, parameter x can be integrated out, leading to the posterior distribution

$$p(\pi \mid y) \propto \prod_{j=1}^{\#\pi} \int_{\mathbb{X}} \prod_{i \in \pi_j} \kappa(y_i; x) \nu_0(\mathrm{d}x; \phi)\, \mu_0(\pi; \psi) = \prod_{j=1}^{\#\pi} L(y; \pi_j, \phi)\, \mu_0(\pi; \psi), \tag{3}$$

where L is the marginal likelihood of cluster π_j.

Even though Model (1) is quite similar to $\mathscr{P}_{[n]}$-partition-based proposals for clustering, the fact that Π is defined over $\mathscr{OP}_{[n]}$, a superset of $\mathscr{P}_{[n]}$, allows us to analyze more features. For example, it is possible to compute posterior distributions related to kernel parameters, e.g., $p(x \mid \pi, y)$, whereas under a $\mathscr{P}_{[n]}$ approach, this is possible through the use of extra procedures.

3 Probability Distributions over Ordered Set Partitions

Having probability distributions over $\mathscr{OP}_{[n]}$ is required in order to explore the performance of Model (1). To the best of our knowledge, there are few approaches working over this space. In collaborative filtering, [9, 10] define particular product partition distributions, and [1] proposes a stochastic process over this space but no concrete distribution over $\mathscr{OP}_{[n]}$ is provided. Therefore, we will derive an $\mathscr{OP}_{[n]}$-valued distribution based on the d-indicator-vector context.

It is interesting to mention that many authors working on mixture models agree on the support of the indicator vector d (see, for example [4, 7, 8]), but their adopted approaches are not defined on it. Mixing weights induce the probability distribution for the indicator vector d, but since their priors are symmetric, the resulting probability distributions degenerate to the class of set partitions $\mathscr{P}_{[n]}$.

Let $d = (d_1, \ldots, d_n)$ be a random vector where each d_i is a discrete random variable. Assume each d_j is independent and identically distributed, so

$$\Pr(d_1 = a_1, \ldots, d_n = a_n) = \prod_{i=1}^{n} \Pr(d_i = a_i).$$

To obtain an $\mathscr{OP}_{[n]}$-valued probability distribution, an appropriate relabeling is required. Denoting by $a^* = (a_1^*, \ldots, a_{k^*}^*)$ the k^* distinct values in $(a_1, \ldots, a_n)$, we need to define a bijection between a^* and the set $[k^*]$. In particular, let $\iota = (\iota_1, \ldots, \iota_n)$ be a permutation of $[n]$ such that $a_{\iota_1} \leq a_{\iota_2} \leq \cdots \leq a_{\iota_n}$. Therefore, the relabeled values, denoted by $a_1', \ldots, a_n'$, should be such that $1 = a_{\iota_1}' \leq a_{\iota_2}' \leq \cdots \leq a_{\iota_n}' = k^*$ for the given ι. The induced ordered set partition $\pi = \pi_1 / \ldots / \pi_{k^*}$ is then

$$\pi_j = \{i : a_i' = j\}, \qquad j = 1, \ldots, k^*.$$

Notice that the clustering of the data does not change neither does the order among groups; we have only removed empty groups.

As a particular example, take the indicator variable d_i geometric distributed, with parameter λ, taking values in $\{1, 2, \ldots\}$, for each $i = 1, \ldots, n$, i.e., $\Pr(d_i = j) = (1 - \lambda)\lambda^{j-1}$. Then

$$\Pr(d_1 = a_1, \ldots, d_n = a_n) \propto \lambda^{\sum_{i=1}^{n} a_i}; \tag{4}$$

the sum in the exponent can be expressed in terms of the unique values a^* as

$$\sum_{i=1}^{n} a_i = \sum_{j=1}^{k^*} a_j^* \,\#\{a_i = a_j^*\},$$

with $\#A$ the cardinality of the set A. After performing the relabeling, the analogous expression for the sum in (4) leads to the $\mathcal{OP}_{[n]}$-valued distribution given by

$$\Pr(\Pi = \pi_1/\ldots/\pi_k) \propto \lambda^{\sum_{j=1}^{k^*} j\,\#\pi_j}. \tag{5}$$

This distribution is not symmetric; unless all the groups have the same size, we will have $\Pr(\Pi = \pi_1/\ldots/\pi_{k^*}) \neq \Pr(\Pi = \pi_{\rho(1)}/\ldots/\pi_{\rho(k^*)})$ for any permutation ρ of $[k]$. Indeed, this distribution assigns higher probabilities to partitions $\pi \in \mathcal{OP}_{[n]}$ such that $\#\pi_1 \geq \#\pi_2 \geq \cdots \geq \#\pi_{k^*}$ (cf. [5]).

The followed approach to build this probability distribution preserves appealing features of $\mathcal{P}_{[n]}$- and $\mathcal{W}_{[n]}$-based models. The relabeling procedure essentially translates a $\mathcal{W}_{[n]}$-valued distribution to an $\mathcal{OP}_{[n]}$-valued one without removing, for example, the model's interpretability under a mixture model framework. Moreover, since the class $\mathcal{OP}_{[n]}$ generalizes $\mathcal{P}_{[n]}$, features of the latter are almost automatically inherited to the former, but it is also possible to incorporate additional group properties in the $\mathcal{OP}_{[n]}$-valued probability distribution. In particular, the symmetry property was relaxed in (5) favoring partitions whose blocks are ordered by size.

4 Sampling Scheme and Inference

In order to simulate from the posterior distribution (2), a Markov chain Monte Carlo (MCMC) scheme is needed. Updating kernel parameters $x_1, \ldots, x_k$ can be done by means of the conditional posterior density

$$p(x_j|x_{-j}, \pi, y) \propto \prod_{i\in\pi_j} \kappa(y_i; x_j)\nu_0(x_j; \theta), \qquad j = 1, \ldots, \#\pi,$$

and simulating from it is straightforward, particularly when κ and ν_0 form a conjugate pair. The second step consists of sampling from the conditional distribution for the partition π. We make use of the indicator vector d for this last step.

Examining the possible changes in the partition π by means of d, suppose $d_i = j$ for some j, meaning that observation y_i belongs to cluster π_j. Thus, π can be modified according to the following cases:

1. Observation y_i is moved to group π_r, that is, $d_i \leftarrow r$. If $r = j$, the partition does not change; otherwise, there are two possibilities. Suppose $\pi = \pi_1/\ldots/\pi_k$, then
   a. $\#\pi_j > 1$, so the new partition π' will contain the same number of groups; only two groups are modified: $\pi'_j = \pi_j \setminus \{i\}$ and $\pi'_r = \pi_r \cup \{i\}$.
   b. $\#\pi_j = 1$, meaning that such a group disappear; the new partition is

$$\pi' = \pi_1/\ldots/\pi_{j-1}/\pi_{j+1}/\ldots/\pi_k,$$

   with $\pi'_r = \pi_r \cup \{i\}$. Note that the kernel parameters also have to be rearranged, so they match their corresponding group, therefore

$$x' = (x_1, \ldots, x_{j-1}, x_{j+1}, \ldots, x_k).$$

2. Observation y_i is moved to a new group. We have similar cases to the previous ones, but the new partition will contain the group $\{i\}$, with associated parameter x^* drawn from the prior ν_0.

An important point to be considered is the position of the new group in the second case, since the order among groups is relevant under an $\mathscr{OP}_{[n]}$-approach. Denote by $\pi^* = \pi_1^*/\ldots/\pi_{k^*}^*$ the partition induced by removing item i as explained as a first step. The updated partition π' can have the new group $\{i\}$ between any couple of groups π_j^* and π_{j+1}^*, $j = 1, \ldots, k^* - 1$, or just before group π_1^*, or just after group $\pi_{k^*}^*$. Thus, there are $k' = k + 1$ different ways to place group $\{i\}$ whenever $\#\pi_j > 1$, and $k' = k - 1$ otherwise. Therefore, the corresponding conditional posterior distribution for d_i, $i = 1, \ldots, n$, is given by

$$p(d_i = j | d_{-i}, x, y) \propto \begin{cases} \kappa(y_i; x_j)\mu_0(\pi_{d_i \leftarrow j}), & j = 1, \ldots, k, \\ \kappa(y_i; x^*)\mu_0(\pi \overset{j-k}{\vee} \{i\}), & j = k+1, \ldots, k+k', \end{cases}$$

where $x^* \sim \nu_0$, $\pi_{d_i \leftarrow j}$ denotes the induced partition obtained when no new group appears, and $\pi \overset{r}{\vee} \{i\}$ is the partition π with item i as a singleton placed just before group r, as already explained, with $r = k'$ meaning the new group is placed after the last group.

4.1 Distribution for the Number of Groups

Apart from the clustering structure, and similar to $\mathscr{P}_{[n]}$-based approaches, it is possible to infer about the number of groups using Model (1). Denote by $\mathscr{OP}_{[n]}^k$ the subset of all ordered set partitions with exactly k groups. Thus, the probability distribution for the number of groups, k^*, can be calculated by marginalizing from the $\mathcal{OP}_{[n]}$-valued distribution of Π as follows:

$$\Pr(k^* = j) = \sum_{\pi \in \mathcal{OP}_{[n]}^j} \Pr(\Pi = \pi), \qquad j = 1, \ldots, n.$$

Furthermore, from the posterior distribution (3), the posterior distribution for k^* is given by

$$p(k^* = j|y) \propto \sum_{\pi \in \mathcal{OP}_{[n]}^j} p(\pi|y).$$

Let $\pi^{(1)}, \ldots, \pi^{(T)}$ be a sample of size T from the posterior distribution (2). The posterior distribution for the number of groups can be computed as

$$p(k^* = j|y) = \frac{1}{T} \sum_{i=1}^{T} \mathbf{1}(\#\pi^{(i)} = j).$$

4.2 Estimated Density

Another estimator of interest from Model (1) is the induced mixture density, which is defined as

$$f(y|\pi, x) = \sum_{j=1}^{\#\pi} \frac{\#\pi_j}{n} \kappa(y; x_j), \qquad \pi \in \mathcal{OP}_{[n]}, \tag{6}$$

where the mixing weights are given by $w_j := \#\pi_j/n$. Furthermore, marginalizing over (π, x), we have

$$f(y) = \sum_{\pi \in \mathcal{OP}_{[n]}} \int_{\mathbb{X}^{\#\pi}} f(y|\pi, x) \nu_0^{\#\pi}(\mathrm{d}x) \mu_0(\pi) = \sum_{\pi \in \mathcal{OP}_{[n]}} \sum_{j=1}^{\#\pi} w_j \int_{\mathbb{X}} \kappa(y; x_j) \nu_0(\mathrm{d}x_j)\, \mu_0(\pi). \tag{7}$$

Therefore, both expressions can be used to compute an estimation for the underlying mixture density of Model (1).

Let $(\pi^{(1)}, x^{(1)}), \ldots, (\pi^{(T)}, x^{(T)})$ be the T sampled values from the posterior distribution (2), where $x^{(t)} = (x_1^{(t)}, \ldots, x_{k^{(t)}}^{(t)})$ with $k^{(t)} = \#\pi^{(t)}$. Then, for every $t = 1, \ldots, T$, the density function (6) is estimated by

$$\hat{f}^{(t)}(y) = \sum_{j=1}^{\#\pi^{(t)}} \frac{\#\pi_j^{(t)}}{n} \kappa(y; x_j^{(t)}),$$

and the marginal density (7) as

$$\hat{f}(y) = \frac{1}{T} \sum_{t=1}^{T} \hat{f}^{(t)}(y).$$

A variant of this latter estimate is obtained when we consider a specific ordered partition, π^*, such as the posterior mode. In this case, the estimated density is given by

$$\hat{f}(y|\pi^*) = \frac{1}{T^*} \sum_{t=1}^{T} \hat{f}^{(t)}(y)\, \mathbf{1}(\pi^{(t)} = \pi^*).$$

with $T^* = \#\{\pi^{(t)} = \pi^* : t = 1, \ldots, T\}$.

5 Numerical Illustrations

Two datasets are used next to explore the performance of the proposed model. The first one is a toy dataset, which allows us to compare simulation results with the analytic ones, whereas the second corresponds to the well-known galaxies dataset.

A Gaussian kernel is assumed to model observations in each group, so $\kappa(y; x) = \mathrm{N}(y; m, 1/v)$, where $x = (m, v)$, with a conjugate prior for x, namely, a Normal-Gamma distribution, thus $m|v \sim \mathrm{N}(m; m_0, c_0/v)$ and $v \sim \mathrm{Ga}(v; a_0, b_0)$. Then, given a sample $y = (y_1, \ldots, y_n)$, the posterior distribution is

$$\mathrm{N}\left(m; \frac{c_0 n \bar{y} + m_0}{c_0 n + 1}, \frac{c_0}{v(c_0 n + 1)}\right) \mathrm{Ga}\left(v; \frac{n}{2} + a_0, \frac{n(\bar{y} - m_0)^2}{2(c_0 n + 1)} + \frac{1}{2}\sum_{i=1}^{n}(y_i - \bar{y})^2 + b_0\right).$$

The toy dataset we will use has been taken from [5], which is

$$-1.521,\ -1.292,\ -0.856,\ -0.104,\ 2.388,\ 3.079,\ 3.312,\ 3.415,\ 3.922,\ 4.194,$$

and is displayed in Fig. 1. The advantage of using a small dataset is that we can compute all the probabilities without numerical errors produced by simulation schemes and compare them with those obtained using the sampling scheme presented previously. Base measure parameters used are $(m_0, c_0, a_0, b_0) = (0, 100, 1, 1)$, and for the partition distribution, its parameter was set to $\lambda = 0.2$. Table 1 presents the highest posterior probabilities for the partitions and the posterior distribution for the number of groups for the simulation, together with the exact results. The posterior modal partition corresponds to a grouping with two clusters, where the first and biggest one is formed by observations $(y_5, \ldots, y_{10})$ and the second cluster by observations $(y_1, \ldots, y_4)$.

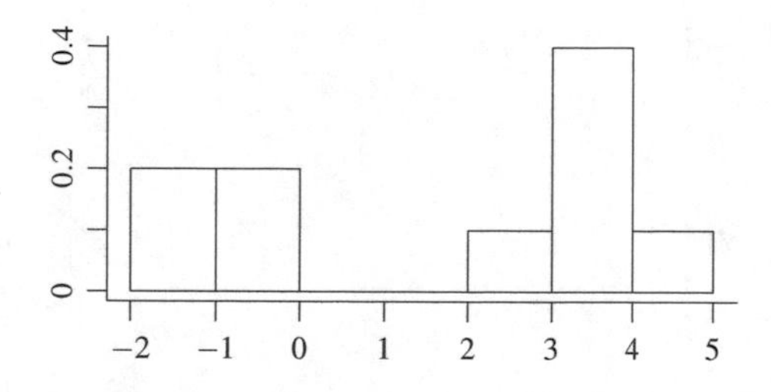

Fig. 1 Histogram of the toy example dataset

Table 1 Posterior probabilities for the toy example dataset. In both cases, the probabilities obtained via the MCMC sampler and by exact computations are given

π	$p(\pi \mid y)$	
	MCMC	Exact
(2, 2, 2, 2, 1, 1, 1, 1, 1, 1)	0.69978	0.69927
(2, 2, 2, 1, 1, 1, 1, 1, 1, 1)	0.03252	0.03552
(2, 2, 2, 3, 1, 1, 1, 1, 1, 1)	0.02984	0.02845
(1, 1, 1, 1, 2, 2, 2, 2, 2, 2)	0.02724	0.02797
(1, 1, 1, 1, 1, 1, 1, 1, 1, 1)	0.02348	0.02018

(a) Highest probabilities for the partitions. Each partition π is shown using its corresponding indicator vector d representation

	$p(k^* \mid y)$	
k^*	MCMC	Exact
1	0.02348	0.02018
2	0.83738	0.84145
3	0.13628	0.13544
4	0.00284	0.00292
5	0.00002	0.00001

(b) Posterior distribution for the number of groups, k^*

Notice that labels, d_i, enable inference about the clustering structure, and also indicate the position of each group in terms of its size: the biggest group is identified with the number 1, the second bigger group with 2, and so on: the ith bigger group will have label i. This is not possible to do under a $\mathscr{P}_{[n]}$ approach.

The second dataset corresponds to the galaxies dataset, widely used in clustering and density estimation applications. Keeping the kernel and prior distribution for its parameters from previous example, base measure parameters used are $(m_0, c_0, a_0, b_0) = (21, 450, 2, 1)$, and for parameter λ two values were used: 0.1 and 0.5. For each specification, a sample of size 20 000 was taken from the MCMC after discarding 10 000, and the posterior modal clustering, density estimation, and posterior distribution for the number of groups are reported in Figs. 2, 3, and Table 2, respectively.

The posterior modal ordered partition appears in Figs. 2 and 3 below the histograms. Similar to Table 1, label 1 identifies to the biggest group, and, in general, label i identifies the ith bigger group. An important fact to be mentioned here is that no additional procedures were needed to order the groups by size. Furthermore, given this partition, the displayed density was computed. On the other hand, our model also allows us to compute the posterior distribution of each kernel parameter, conditioned on some partition. For the case $\lambda = 0.1$, these distributions are presented in Fig. 4 when we condition on the posterior modal partition. Additionally, mixing

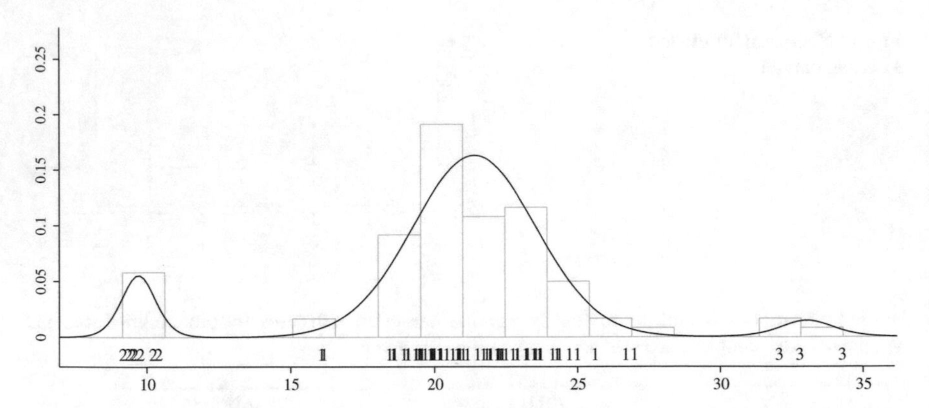

Fig. 2 Estimated density for the galaxies dataset, using $\lambda = 0.1$. Numbers below the histogram represent the posterior modal partition; their positions over the x-axis correspond to their associated observed values

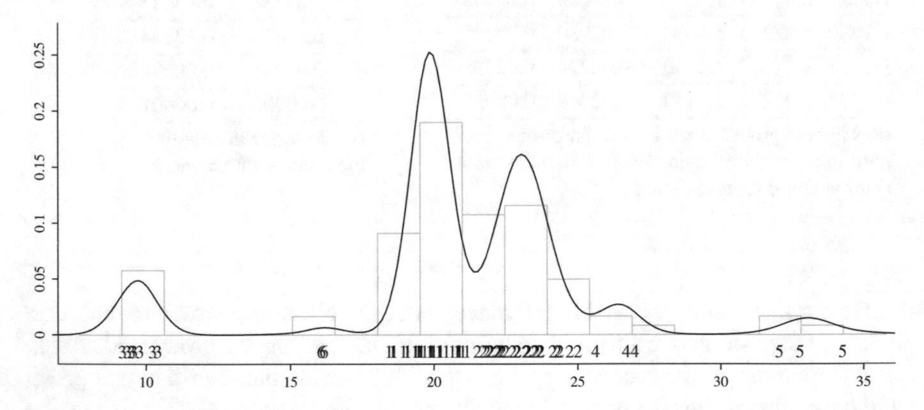

Fig. 3 Estimated density for the galaxies dataset, using $\lambda = 0.5$. Numbers below the histogram represent the posterior modal partition; their positions over the x-axis correspond to their associated observed values

weights can be estimated, which are $(\hat{w}_1, \hat{w}_2, \hat{w}_3) = (72/82, 7/82, 3/82)$ under this scenario; these are quite similar to the ones obtained by [5].

We can see that parameter λ plays a role in the clustering structure. When $\lambda = 0.1$, the dataset is partitioned into three groups, and the distribution for the number of groups is also concentrated around this value, but if λ increases, the biggest group is disaggregated in such a way the modal partition has six groups, and the distribution for the number of groups also changes accordingly.

Table 2 Posterior distribution for the number of groups, k^*, for each value of λ, obtained for the galaxies dataset

	k^*							
λ	3	4	5	6	7	8	9	10
0.1	0.92000	0.07965	0.00035	—	—	—	—	—
0.5	—	0.00185	0.04800	0.42395	0.40225	0.11135	0.01220	0.00040

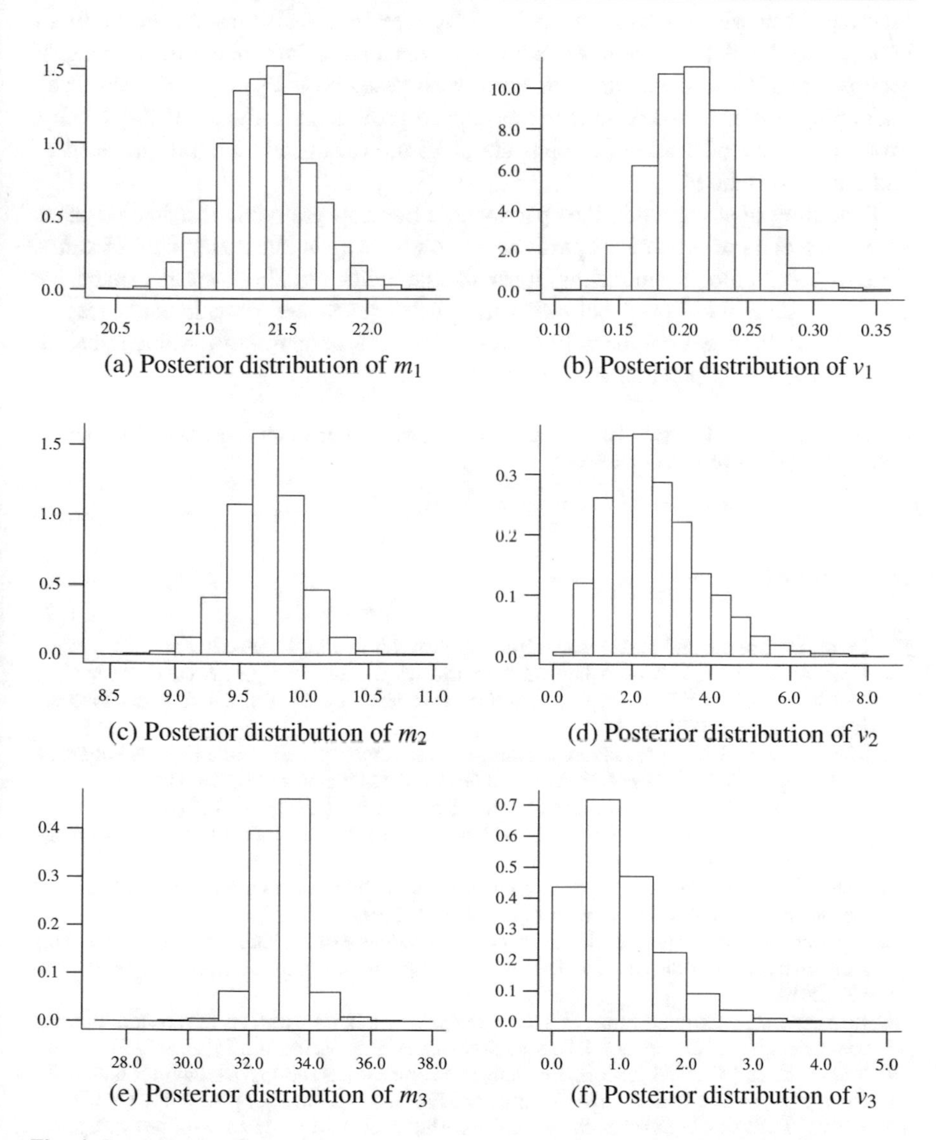

(a) Posterior distribution of m_1 (b) Posterior distribution of v_1

(c) Posterior distribution of m_2 (d) Posterior distribution of v_2

(e) Posterior distribution of m_3 (f) Posterior distribution of v_3

Fig. 4 Posterior distribution of kernel parameters given the posterior modal clustering for the galaxies dataset when $\lambda = 0.1$

6 Discussion and Ongoing Work

We have presented a novel approach to perform clustering based on random partitions by relaxing the symmetry property of $\mathscr{P}_{[n]}$-valued distributions. There are many implications of this change. On the one hand, it is possible to model additional partition features. Here, our approach includes the size of each group in the partition, favoring those whose groups are ordered by size from left to right. On the other hand, since this is still a partition approach, we can also infer about the number of groups. In addition, since our approach is also based on a $\mathscr{W}_{[n]}$ context, there is an underlying mixture model, so it is possible to provide an estimate of the density. Moreover, our model allows to infer about kernel parameters without the need of additional procedures.

Regarding ongoing work, the space where our nonsymmetric distribution takes values, the class of ordered set partitions, is on its own an interesting field of study. With respect to mixture modeling, it seems that, when working over this space, we can better understand the label-switching phenomenon and provide some way to overcome it. From a more theoretical perspective, the study of $\mathscr{OP}_{[n]}$-valued random partitions seems to be a new subject for research.

Acknowledgements I would like to thank two anonymous referees for many helpful comments made on a previous version of the paper.

References

1. Crane, H.: The cut-and-paste process. Ann. Probab. **42**(5), 1952–1979 (2014)
2. Lijoi, A., Prünster, I.: Models beyond the Dirichlet process. In: Hjort, N.L., Holmes, C.C., Müller, P., Walker, S.G. (eds.) Bayesian Nonparametrics, pp. 80–136. Cambridge University Press, Cambridge (2010)
3. Martínez, A.F.: Usages of random combinatorial structures in statistics; a Bayesian nonparametric approach. Ph.D. thesis, Universidad Nacional Autónoma de México (2015)
4. McCullagh, P., Yang, J.: How many clusters? Bayesian Anal. **3**(1), 101–120 (2008)
5. Mena, R.H., Walker, S.G.: On the Bayesian mixture model and identifiability. J. Comput. Graph. Stat. **24**(4), 1155–1169 (2015)
6. Nobile, A., Fearnside, A.T.: Bayesian finite mixtures with an unknown number of components: the allocation sampler. Stat. Comput. **17**, 147–162 (2007)
7. Papastamoulis, P., Iliopoulos, G.: An artificial allocations based solution to the label switching problem in Bayesian analysis of mixtures of distributions. J. Comput. Graph. Stat. **19**(2), 313–331 (2010)
8. Richardson, S., Green, P.J.: On Bayesian analysis of mixtures with an unknown number of components (with discussion). J. R. Stat. Soc: Ser. B (Stat. Methodol.) **59**(4), 731–792 (1997)
9. Tran, T., Phung, D., Venkatesh, S.: Learning from ordered sets and applications in collaborative ranking. In: JMLR: Workshop and Conference Proceedings, vol. 25, pp. 427–442 (2012)
10. Truyen, T., Phung, D., Venkatesh, S.: Probabilistic models over ordered partitions with applications in document ranking and collaborative filtering. In: Proceedings of the 2011 SIAM International Conference on Data Mining, pp. 426–437 (2011)

A Flexible Replication-Based Classification Approach for Parkinson's Disease Detection by Using Voice Recordings

Lizbeth Naranjo, Ruth Fuentes-García and Carlos J. Pérez

Abstract Detecting Parkinson's disease (PD) by using a noninvasive low-cost tool based on acoustic features automatically extracted from voice recordings has become a topic of interest. A two-stage classification approach has been developed to differentiate PD subjects from healthy people by using acoustic features obtained from replicated voice recordings. The proposed hierarchical model has been specifically developed to handle replicated data and considers a dimensional reduction of the feature space as well as the use of mixtures of normal distributions to describe the latent variables in the second order of hierarchy. The approach has been applied to a database of acoustic features obtained from 40 PD subjects and 40 healthy controls, improving results compared to previous models.

Keywords Bayesian binary hierarchical model · Common principal components · Mixtures of normal distributions · Parkinson's disease · Replicated measurements · Voice recordings

1 Introduction

Parkinson's disease (PD) is a progressive nervous system disorder that mainly affects movement. PD is the second most common neurodegenerative disorder after Alzheimer's disease, affecting an estimated 7–10 million people worldwide accord-

L. Naranjo · R. Fuentes-García (✉)
Departamento de Matemáticas, Facultad de Ciencias, Universidad Nacional Autónoma de México, 04510 Mexico City, Mexico
e-mail: rfuentes@ciencias.unam.mx

L. Naranjo
e-mail: lizbethna@ciencias.unam.mx

C. J. Pérez
Departamento de Matemáticas, Facultad de Veterinaria, Universidad de Extremadura, 10003 Cáceres, Spain
e-mail: carper@unex.es

I. Antoniano-Villalobos et al. (eds.), *Selected Contributions on Statistics and Data Science in Latin America*, Springer Proceedings in Mathematics & Statistics 301,
https://doi.org/10.1007/978-3-030-31551-1_7

ing to the Parkinson's Disease Foundation[1] [30]. The symptoms are tremor, slow movements, muscular stiffness, impaired posture and balance, loss of automatic movements, and communication problems, among the most important ones. Symptoms start gradually and become disabling when time progresses. Up to now, there is no cure but medications and, in some cases, surgery may significantly improve the symptoms.

Most people with PD, at some stage, experience problems with their voice. Hypokinetic dysarthria is the most common communication disorder, causing rigidity and slowness of the systems of communication including swallowing, breathing, and speech. These problems in the voice production usually lead to social isolation. Vocal impairment can be one of the earliest indicators of PD [9].

In recent years, statistical models using features automatically extracted from voice recordings have shown to provide an effective, noninvasive mode of discriminating PD subjects from healthy people [10, 18, 29]. This has been motivated by the increased interest in building predictive telediagnosis models that may help in the early stage PD detection. Reference [26] presents a literature review on PD diagnosis through features extracted from speech. A recent comparative study of some machine learning approaches for PD detection can be found in [19].

In this context, replicated voice recordings have been considered, leading to within-subject variability, since consecutive voice recordings from the same person at a concrete time are not identical due to technology imperfections and biological variability. However, most authors have used the corresponding replicated features as if they were independent, obviating their dependent nature. Reference [21] presented a logistic-based classification approach for PD detection that takes into account the underlying within-subject dependence of the recordings. Reference [15] also took into account the within-subject variability with an approach based on the probit regression. Later, [16] considered this framework in a variable selection context. The approaches in [15, 16] were tested on a database specifically designed and collected for this purpose that is composed of three voice recording replications of the sustained /a/ phonation for 40 subjects with PD and 40 healthy people.

The presence of high correlations between variables results in a multicollinearity problem, causing estimates to be unstable and with possible bias. It also may lead to large standard error for parameter estimates and/or parameter estimates with opposite signs than those expected. Some solutions are variable selection [17, 24], regularization [13, 20], or dimensional reduction [11].

The approaches that consider the within-subject variability may be improved by using a flexible measurement error for the unknown covariate vectors. In specific, latent variables can be introduced to model the features automatically extracted from the speech recordings with unimodal or multimodal patterns using mixtures of normal distributions [2]. This may result in higher classification accuracies.

In this paper, a new classification approach has been developed to differentiate people with Parkinson's disease from healthy subjects by using acoustic features obtained from replicated voice recordings. The approach extends those in [15, 16]

[1]https://parkinsonsnewstoday.com/parkinsons-disease-statistics/.

by considering a flexible measurement error for the unknown covariate vectors. Multicollinearity problems have been avoided by using common principal component analysis (CPCA). Since the extracted features display a correlation structure that is stable throughout the replications, CPCA is appropriate to handle replicated measurement. The approach has been applied to several scenarios for comparison purpose.

The remainder of this paper is as follows. Section 1 gives a brief introduction. The motivating problem is described in Sect. 2. Section 3 shows the proposed methodology. In Sect. 4, the Bayesian analysis of the proposed model is developed. Section 5 presents the experimental results. Finally, Sect. 6 shows the conclusion on some specific points of the proposed methodology and its advantages.

2 Motivating Problem

PD detection based on acoustic features automatically extracted from voice recordings is a topic of current interest in the scientific literature. Reference [12] presented one of the most widely extended databases on this topic based on phonations of sustained /a/. Signal processing algorithms were used to extract acoustic features from the voice recordings. Many of these processing algorithms have slight differences, providing highly correlated features. For instance, shimmer measures, which are related to the amplitude variation of the individual pitch periods of the fundamental frequency, have some related definitions as, for example, relative shimmer defined by

$$\text{Relative shimmer} = \frac{\frac{1}{N-1}\sum_{i=1}^{N-1} |A_i - A_{i+1}|}{\frac{1}{N}\sum_{i=1}^{N} A_i},$$

where A_i is maximum amplitude of the pitch periods, whereas APQ3 is defined by

$$\text{APQ3} = \frac{\frac{1}{N-2}\sum_{i=2}^{N-1} \left| A_i - \frac{A_{i-1}+A_i+A_{i+1}}{3} \right|}{\frac{1}{N}\sum_{i=1}^{N} A_i}.$$

Another four shimmer-related definitions have been considered in the scientific literature, leading to highly pairwise correlated features. The same happens with other groups of related features. This motivates the need for reducing the dimensionality of the feature space by keeping the main information useful to differentiate PD subjects from healthy ones. Even more, this dimensionality reduction must be performed in a replication-based context.

In this paper, data from the experiment conducted by [15] is used. It is based on acoustic features extracted from three voice recording replications of the sustained /a/ phonation for each one of the 80 subjects (40 with PD and 40 healthy subjects). Each voice recording was processed to provide 44 acoustic features per voice recording. The extracted features were grouped according to whether they had related formulation or not. This led to eight groups, four of them having one only feature, i.e.,

$\mathscr{G}_1$ Pitch local perturbation measures: relative jitter, absolute jitter, relative average perturbation (RAP), and pitch perturbation quotient (PPQ).
$\mathscr{G}_2$ Amplitude perturbation measures: local shimmer, shimmer in dB, 3-point amplitude perturbation quotient (APQ3), 5-point amplitude perturbation quotient (APQ5), and 11-point amplitude perturbation quotient (APQ11).
$\mathscr{G}_3$ Harmonic-to-noise ratio measures: harmonic-to-noise ratio in the frequency band 0–500 Hz (HNR05), in 0–1500 Hz (HNR15), in 0–2500 Hz (HNR25), in 0–3500 Hz (HNR35), and in 0–3800 Hz (HNR38).
$\mathscr{G}_4$ Mel frequency cepstral coefficient-based spectral measures of order 0 to 12 (MFCC0, MFCC1,..., MFCC12) and their derivatives (Delta0, Delta1,..., Delta12).
$\mathscr{G}_5$ Recurrence period density entropy (RPDE).
$\mathscr{G}_6$ Detrended fluctuation analysis (DFA).
$\mathscr{G}_7$ Pitch period entropy (PPE).
$\mathscr{G}_8$ Glottal-to-noise excitation ratio (GNE).

The dataset can be downloaded from UCI Machine Learning repository https://archive.ics.uci.edu/ml/datasets/Parkinson+Dataset+with+replicated+acoustic+features+.

In order to avoid a multicollinearity problem, [16] selected only one feature per group. Other scenarios for variable reduction can be applied for the first stage, including methods related to common principal component which is appropriate since the covariance structure is very similar in each of the three replications. In a second stage, these variables feed a flexible classification approach that takes into account mixtures of normal distributions. Therefore, the aim is to reduce the number of variables with reduction techniques and a classification approach that is compatible with the considered replication-based framework while keeping a high discrimination power.

3 Methodology

In this section, the methodology used in this paper is described. First, the common principal component analysis is presented. Second, the hierarchical model to deal with binary response variables is shown. Finally, the use of finite mixture model for the latent variables to deal with the replications of the covariates is described.

3.1 Common Principal Component Analysis

The common principal component (CPC) analysis is a generalization of principal component analysis (PCA) that allows to reduce the dimension at the same time that deals with replications. For more details, see [6–8, 11].

Suppose there are observations from a random $K \times 1$ vector $\boldsymbol{u}$ and there are J distinct replications, and that the covariance matrix for the jth replication is Σ_j, $j = 1, \ldots, J$. The presence of CPC is defined by the hypothesis that multiple datasets share common components, this means that there is an orthogonal matrix A that simultaneously diagonalizes all the Σ_j so that,

$$A' \Sigma_j A = \Lambda_j,$$

where Λ_j is a diagonal matrix for each $j = 1, \ldots, J$. The kth column of A gives the coefficients of the kth CPC, and the elements of Λ_j give the variances of these CPC's for the jth replication. The simultaneously transformed variables $\boldsymbol{x}_j = A'\boldsymbol{u}_j$ are called *common principal components*, which are orthogonal. This way of deriving components keeps the correlation structure of the replications.

Note that $\boldsymbol{u}_j$'s denote all the covariates in the database and $\boldsymbol{x}_j$'s denote the CPC after applying the CPC analysis. $\boldsymbol{x}_j$'s are used in Sect. 3.2 as covariates inside a hierarchical model.

3.2 Binary Response Model

The binary response model constitutes the first level of the hierarchical model. Suppose that n independent binary random variables $Y_1, \ldots, Y_n$ are observed, where Y_i is distributed as a Bernoulli with success probability $\text{p}(Y_i = 1) = \theta_i$, $i = 1, \ldots, n$. The probabilities θ_i are related to two sets of covariates $\boldsymbol{x}_i$ and z_i, where $\boldsymbol{x}_i = (\boldsymbol{x}_{i1}, \ldots, \boldsymbol{x}_{iK})'$ is a $K \times J$ matrix of a set of K covariates which have been measured with J replicates, and $z_i = (z_{i1}, \ldots, z_{iH})'$ is a vector of a set of H covariates which are exactly known. Let $\boldsymbol{x}_{ij} = (x_{i1j}, \ldots, x_{iKj})'$ be the jth replication of the unknown covariate vector $\boldsymbol{w}_i = (w_{i1}, \ldots, w_{iK})'$, $j = 1, \ldots, J$, and assume that they have a linear relationship represented by an additive measurement error model (see, e.g., [1, 3]). This implies that instead of the covariates $\boldsymbol{w}_i$, their replicates $\boldsymbol{x}_{ij}$ are observed, i.e., the $\boldsymbol{x}_{ij}$ are the surrogates of $\boldsymbol{w}_i$. The parameters θ_i are related to $\boldsymbol{x}_{ij}$ and z_i through the following hierarchical model, similar to the ones proposed in [15, 16]:

$$\begin{aligned} Y_i &\sim \text{Bernoulli}(\theta_i), \\ \Psi^{-1}(\theta_i) &= \boldsymbol{w}_i'\boldsymbol{\beta} + \boldsymbol{z}_i'\boldsymbol{\gamma}, \\ x_{ijk} &= w_{ik} + \varepsilon_{ijk}, \\ \varepsilon_{ijk} &\sim \text{Normal}(0, \sigma_k), \end{aligned}$$

for $i = 1, \ldots, n,\ j = 1, \ldots, J$ and $k = 1, \ldots, K$, where $\boldsymbol{\beta}$ and $\boldsymbol{\gamma}$ are vectors of unknown parameters, of dimensions K and H, respectively. $\Psi^{-1}(\cdot)$ is a known non-decreasing function ranging between 0 and 1, usually it is the inverse of the cumulative distribution function (cdf) of normal or logistic distributions. The parameters σ_k's are precisions. The error ε_{ijk} is independent of w_{ik}, so x_{ijk} can be considered as a surrogate of w_{ik}. The surrogate $\boldsymbol{x}_{ij}$ is assumed to be an error-prone measurement of the true $\boldsymbol{w}_i$.

3.3 Finite Mixture Model

An approach that extends the ones proposed in [15, 16] is developed here. The aim is to include a flexible measurement error for the unknown covariate vectors $\boldsymbol{w}_i$. Specifically, mixtures of normal distributions to address the replications in the covariates are used.

The measured covariates show multimodal patterns, so mixtures of g_k fixed Gaussian distributions [2] are considered in the second order of hierarchy to model the latent variables w_{ik}, for $i = 1, \ldots, n$ and $k = 1, \ldots, K$, i.e.,

$$\begin{aligned} w_{ik} &\sim \mathscr{F}_{W_k}(\boldsymbol{\mu}_k, \boldsymbol{\tau}_k, \boldsymbol{q}_k), \\ f_{W_k}(w_{ik}|\boldsymbol{\mu}_k, \boldsymbol{\tau}_k, \boldsymbol{q}_k) &= \sum_{l=1}^{g_k} q_{kl}\text{p}(w_{ik}|\mu_{kl}, \tau_{kl}), \end{aligned}$$

where $0 \leq q_{kl} \leq 1,\ \sum_{l=1}^{g_k} q_{kl} = 1,\ \boldsymbol{q}_k = (q_{k1}, \ldots, q_{kg_k}),\ \boldsymbol{\mu}_k = (\mu_{k1}, \ldots, \mu_{kg_k})$, $\boldsymbol{\tau}_k = (\tau_{k1}, \ldots, \tau_{kg_k})$, and $\text{p}(w_{ik}|\mu_{kl}, \tau_{kl})$ denotes a Gaussian distribution with mean μ_{kl} and precision τ_{kl}.

Note that in any mixture model, if all the g_k components belong to the same parametric family as in this case, then $f_{W_k}(w_{ik}|\boldsymbol{\mu}_k, \boldsymbol{\tau}_k, \boldsymbol{q}_k)$ is invariant under the $g_k!$ permutations of the component labels in the parameter space. This is known as *label switching* which causes identifiability problems. These problems are handled by imposing an *identifiability constraint* on the parameter space, i.e., $\mu_{k1} < \mu_{k2} < \cdots < \mu_{kg_k}$

In order to generate the latent variables, consider a vector of categorical random variables d_{ik} for $i = 1, \ldots, n$ that takes value in $1, 2, \ldots, g_k$. Regarded as allocation variables for the observations, they are assumed to be independent draws from the distributions,

$$\mathrm{p}(d_{ik} = l|\boldsymbol{\mu}_k, \boldsymbol{\tau}_k, \boldsymbol{q}_k) = q_{kl},$$

for $l = 1, 2, \ldots, g_k$. Note that $\mathrm{p}(d_{ik} = l|\boldsymbol{\mu}_k, \boldsymbol{\tau}_k, \boldsymbol{q}_k)$ does not depend on $\boldsymbol{\mu}_k$ or $\boldsymbol{\tau}_k$, then $\mathrm{p}(d_{ik} = l|\boldsymbol{\mu}_k, \boldsymbol{\tau}_k, \boldsymbol{q}_k) = \mathrm{p}(d_{ik} = l|\boldsymbol{q}_k)$.

Conditional on the $d_{ik} = l$, the density of w_{ik} is given by $\mathrm{p}(w_{ik}|\mu_{kl}, \tau_{kl})$. The vector $\boldsymbol{d}_k = (d_{1k}, d_{2k}, \ldots, d_{nk})'$ is frequently called the *missing data* part of the sample and is integrated out in many schemes. Let $\mathrm{p}(\cdot)$ denote the corresponding density function for the g_k-component mixture. The unknown parameters $(\boldsymbol{\mu}_k, \boldsymbol{\tau}_k, \boldsymbol{q}_k)$ are drawn from a set of appropriate prior distributions. Explicitly considering the allocation variables, the density of the joint distribution of all variables can be written as

$$\begin{aligned}&\mathrm{p}(\boldsymbol{w}_k, \boldsymbol{\mu}_k, \boldsymbol{\tau}_k, \boldsymbol{d}_k, \boldsymbol{q}_k|g_k) \qquad (1)\\ &\quad= \mathrm{p}(\boldsymbol{w}_k|\boldsymbol{\mu}_k, \boldsymbol{\tau}_k, \boldsymbol{d}_k, \boldsymbol{q}_k, g_k)\mathrm{p}(\boldsymbol{\mu}_k, \boldsymbol{\tau}_k|\boldsymbol{d}_k, \boldsymbol{q}_k, g_k)\mathrm{p}(\boldsymbol{d}_k|\boldsymbol{q}_k, g_k)\mathrm{p}(\boldsymbol{q}_k|g_k),\end{aligned}$$

where $\mathrm{p}(\cdot|\cdot)$ is used to denote generic conditional distributions.

The posterior quantities of interest can then be approximated by numerical methods as the EM algorithm [14] or Markov chain Monte Carlo (MCMC) methods [4].

4 Bayesian Analysis

In this section, the prior distributions are presented and the posterior distribution is derived.

4.1 Prior Distributions

A usual approach for linear models assumes normal distributions for the regression parameters,

$$\begin{aligned}\beta_k &\sim \mathrm{Normal}(b_k, B_k),\\ \gamma_h &\sim \mathrm{Normal}(c_h, C_h),\end{aligned}$$

where $\boldsymbol{b} = (b_1, \ldots, b_K)$, $\boldsymbol{B} = (B_1, \ldots, B_K)$, $\boldsymbol{c} = (c_1, \ldots, c_H)$, and $\boldsymbol{C} = (C_1, \ldots, C_H)$ are fixed values.

For precisions σ_k, Gamma distributions are considered,

$$\sigma_k \sim \mathrm{Gamma}(s_k, r_k),$$

where s_k and r_k are the shape and rate parameters, respectively.

In a finite mixture model context it is not possible to consider fully non-informative or reference prior distributions to obtain proper posterior distributions (see, e.g., [5, 25]), since there is always the possibility that no observations are allocated to one or more components, then informative prior distributions must be considered. We used weakly informative prior distributions to ensure good mixing of the MCMC algorithm (see, e.g., [2, 23]), in specific, the following informative prior distributions are used for the parameters of the mixture:

$$\begin{aligned} \boldsymbol{q}_k &\sim \text{Dirichlet}\,(\boldsymbol{\alpha})\,, \\ \mu_{kl} &\sim \text{Normal}\left(m_k,\, 1/R_k^2\right), \\ \tau_{kl} &\sim \text{Gamma}\,(a_k, \lambda_k)\,, \\ \lambda_k &\sim \text{Gamma}\left(e_k,\, \kappa_k/R_k^2\right), \end{aligned}$$

where m_k is the median of the observed range of x_{ijk}'s, R_k is the length of the interval (the difference between the maximum and minimum of the range). By using m_k and R_k in the hyperparameters of the prior distributions, prior information can be translated into a likely range for the variable spread. For μ_{kl}, its distribution corresponds to the observed range. Moreover, introducing an additional hierarchical level in the prior distribution, allowing λ_k to be random, it is less restrictive, without being informative about their absolute size of the τ_k. See [23] for details.

The following prior distribution is used for $\boldsymbol{\mu}_k$:

$$\mathrm{p}(\boldsymbol{\mu}_k) \propto \mathrm{p}(\mu_{k1})\mathrm{p}(\mu_{k2}|\mu_{k1})\mathrm{p}(\mu_{k3}|\mu_{k2}) \cdots \mathrm{p}(\mu_{km}|\mu_{k,m-1}),$$

where

$$\begin{aligned} \mu_{k1} &\sim \text{Normal}(m_k,\, 1/R_k^2), \\ \mu_{kl}|\mu_{k,l-1} &\sim \text{Normal}(m_k,\, 1/R_k^2)\mathrm{I}[\mu_{kl} > \mu_{k,l-1}], \end{aligned}$$

for $l = 2, \ldots, g_k$.

Note that as hyperparameters of the prior distributions are based on the data, the proposed Bayesian model becomes an Empirical Bayesian model.

4.2 Exploring the Posterior Distribution

Under the hierarchical structure of the model defined in Sects. 3.2 and 3.3, the likelihood function considering the observed and the latent variables is given by

$$
\begin{aligned}
&\mathscr{L}(\boldsymbol{\beta}, \boldsymbol{\gamma}, \boldsymbol{\sigma}, \boldsymbol{\mu}, \boldsymbol{\tau}, \boldsymbol{d}, \boldsymbol{q} \mid \boldsymbol{y}, \boldsymbol{x}, \boldsymbol{z}, \boldsymbol{w}, \boldsymbol{g}) \\
&= \mathrm{p}(\boldsymbol{y}|\boldsymbol{z}, \boldsymbol{w}, \boldsymbol{\beta}, \boldsymbol{\gamma})\mathrm{p}(\boldsymbol{x}|\boldsymbol{w}, \boldsymbol{\sigma})\mathrm{p}(\boldsymbol{w}|\boldsymbol{\mu}, \boldsymbol{\tau}, \boldsymbol{d}, \boldsymbol{q}, \boldsymbol{g}) \\
&= \prod_{i=1}^{n} \left\{ \mathrm{p}(y_i|z_i, \boldsymbol{w}_i, \boldsymbol{\beta}, \boldsymbol{\gamma}) \left[\prod_{j=1}^{J}\prod_{k=1}^{K} \mathrm{p}(x_{ijk}|w_{ik}, \sigma_k) \right] \left[\prod_{k=1}^{K} \mathrm{p}(w_{ik}|\boldsymbol{\mu}_k, \boldsymbol{\tau}_k, \boldsymbol{d}_k, \boldsymbol{q}_k, g_k) \right] \right\}.
\end{aligned} \tag{2}
$$

The joint posterior distribution of the unobservables latent variables $\boldsymbol{w}$ and $\boldsymbol{d}$ and the parameters $\boldsymbol{\beta}$, $\boldsymbol{\gamma}$, $\boldsymbol{\sigma}$, $\boldsymbol{\mu}$, $\boldsymbol{\tau}$, and $\boldsymbol{q}$ is obtained by using the likelihood function (2) and the prior distributions defined in Sect. 4.1, and it is given by

$$
\begin{aligned}
&\pi(\boldsymbol{\beta}, \boldsymbol{\gamma}, \boldsymbol{\sigma}, \boldsymbol{\mu}, \boldsymbol{\tau}, \boldsymbol{d}, \boldsymbol{q} \mid \boldsymbol{y}, \boldsymbol{x}, \boldsymbol{z}, \boldsymbol{w}, \boldsymbol{g}) \\
&\propto \mathscr{L}(\boldsymbol{\beta}, \boldsymbol{\gamma}, \boldsymbol{\sigma}, \boldsymbol{\mu}, \boldsymbol{\tau}, \boldsymbol{d}, \boldsymbol{q} \mid \boldsymbol{y}, \boldsymbol{x}, \boldsymbol{z}, \boldsymbol{w}, \boldsymbol{g})\mathrm{p}(\boldsymbol{\beta})\mathrm{p}(\boldsymbol{\gamma})\mathrm{p}(\boldsymbol{\sigma}) \\
&\times \prod_{k=1}^{K} \left[\mathrm{p}(\boldsymbol{\mu}_k, \boldsymbol{\tau}_k|\boldsymbol{d}_k, \boldsymbol{q}_k, g_k)\mathrm{p}(\boldsymbol{d}_k|\boldsymbol{q}_k, g_k)\mathrm{p}(\boldsymbol{q}_k|g_k) \right].
\end{aligned} \tag{3}
$$

The algorithm has been implemented in JAGS (http://mcmc-jags.sourceforge.net/). R and JAGS codes are available in GitHub repository https://github.com/lizbethna/ParkinsonReplicationClassification.

5 Results

This section presents both the experimental settings and the experimental results obtained by applying the proposed methodology. Besides, the results are compared with others published in the statistical literature.

5.1 Experimental Settings

The dataset considered here is the same as in [15, 16], so the performance is comparable. The acoustic variables have been individually normalized to have mean 0 and standard deviation 1, and the variable sex Z takes values $z = 0$ for men and $z = 1$ for women. The response variable Y takes values $y = 0$ for healthy subjects and $y = 1$ for people with PD.

In order to avoid multicollinearity problems, two scenarios with variable reduction based on CPCs have been considered. Specifically, the first scenario (CPCs per group) considers CPCs calculated independently for each one of the first four groups, i.e., for $\mathscr{G}_g$, $g = 1, 2, 3, 4$, whereas the untransformed features of the remaining groups ($\mathscr{G}_g$, $g = 5, 6, 7, 8$) are considered. From the groups $\mathscr{G}_1$, $\mathscr{G}_2$, and $\mathscr{G}_3$, one CPC per group is obtained providing 95%, 98%, and 98% of the total variability, respectively. For $\mathscr{G}_4$, three CPCs covers 81% of the total variability. Note that for the groups from

$\mathcal{G}_1$ to $\mathcal{G}_8$, K is equal to 4, 5, 5, 26, 1, 1, 1 and 1, respectively. This makes a total of 44 acoustic features that now have been reduced to $K = 10$. The second scenario (jointly selected CPCs) considers CPCs jointly extracted from the first four groups and the only feature from each one of the remaining groups. In this case, four CPCs have been selected and they cover 83% of the total variability. Now, the dimensionality reduction is even greater, leading to a total of $K = 8$ variables. For comparative purpose, four scenarios are considered: (A) Approach [15], (B) Approach [16], (C) CPCs per group, and (D) Jointly selected CPCs.

The variables (CPCs and acoustic features) in scenarios (C) and (D) are used in the proposed approach defined in Sects. 3.2 and 3.3. The number of components (modes) of the finite mixtures of normal distributions for each variable has been chosen separately, independently of the other variables. For each variable, several criteria have been used to choose the number of components. First, as a test, each variable or CPC, denoted by x_{ijk}, is fitted by using the specifications in Sect. 3.3 and considering that it could be modeled by one, two, or three components. Second, some of these options were discarded, mainly those where it was not clear that more that one component was necessary. This was performed either when the distribution of the variable is unimodal, or where problems of lack of identifiability of the components arise, or some label switching problems appeared in spite of the identifiability constraints or the informative prior distributions used. Finally, some criteria were used to choose the best option: the deviance information criterion (DIC) [28], the penalized loss function criterion [22], the mean absolute error (MAE), the mean relative error (MRE), and the root mean squared error (RMSE). The lower they are, the better the fit is.

For the approach (C), the number of modes for each group was the following: 1 mode for $\mathcal{G}_1$, 3 modes for $\mathcal{G}_2$, 1 mode for $\mathcal{G}_3$, 1 mode for the CPC1 of $\mathcal{G}_4$, 2 modes for the CPC2 of $\mathcal{G}_4$, 1 mode for the CPC3 of $\mathcal{G}_4$, 1 mode for $\mathcal{G}_5$, 1 mode for $\mathcal{G}_6$, 3 modes for $\mathcal{G}_7$, and 3 modes for $\mathcal{G}_8$. For the approach (D), the number of modes for each group was 1, 2, 1, and 1 modes, which were used for the CPC1, CPC2, CPC3, and CPC4 jointly extracted from the groups $\mathcal{G}_1, \ldots, \mathcal{G}_4$; 1 mode for $\mathcal{G}_5$, 1 mode for $\mathcal{G}_6$, 3 modes for $\mathcal{G}_7$, and 3 modes for $\mathcal{G}_8$.

The MCMC sampling approach defined in Sect. 4 is applied with the following prior distributions: $\beta_k \sim \text{Normal}(0, 0.01)$, for $k = 1, \ldots, K$, $\gamma \sim \text{Normal}(0, 0.01)$, $\sigma_k \sim \text{Gamma}(0.01, 0.01)$, $\boldsymbol{q}_k \sim \text{Dirichlet}(\boldsymbol{\alpha})$, $\mu_{kl} \sim \text{Normal}\left(m_k, 1/R_k^2\right)$, $\tau_{kl} \sim \text{Gamma}(2, \lambda_k)$, $\lambda_k \sim \text{Gamma}\left(0.5, 10/R_k^2\right)$, where m_k is the midpoint of the observed range of x_{ijk}'s, R_k is the length of the interval. Moreover, only three components have been considered, for which $\boldsymbol{\alpha} = (2, 0, 0)$ if $g_k = 1$, $\boldsymbol{\alpha} = (2, 2, 0)$ if $g_k = 2$, and $\boldsymbol{\alpha} = (2, 2, 2)$ if $g_k = 3$.

A total of 30,000 iterations with a burn-in of 10,000 and a thinning period of 10 generated values are used, providing a sample of length 2,000. With these specifications, the chain generated using the MCMC sampling algorithm seems to have converged. BOA package [27] was used to perform the convergence analysis. The chains require a long burn-in period and the previous specifications are enough to provide evidence of convergence for all parameters in both scenarios.

Table 1 Accuracy metrics for approaches A to D with the considered stratified cross-validation framework. Means (standard deviations)

Criteria	Approach A	Approach B	Approach C	Approach D
Accuracy rate	0.752 (0.086)	0.779 (0.080)	0.836 (0.117)	0.822 (0.120)
Sensitivity	0.718 (0.132)	0.765 (0.135)	0.827 (0.140)	0.821 (0.132)
Specificity	0.786 (0.135)	0.792 (0.150)	0.847 (0.107)	0.837 (0.100)
Precision	0.785 (0.118)	0.806 (0.115)	0.832 (0.070)	0.822 (0.066)
AUC-ROC	0.860 (0.070)	0.879 (0.067)	0.901 (0.059)	0.895 (0.064)

Predictive probabilities are obtained for each subject and a confusion matrix is built to obtain the accuracy metrics. The metrics are accuracy rate ((TP+TN)/n), sensitivity (TP/(TP+FN)), specificity (TN/(TN+FP)), and precision (TP/(TP+FP)). Moreover, the area under the curve of receiver operating characteristic (AUC-ROC) is presented. A stratified cross-validation framework is considered. Specifically, the dataset is randomly split into a training subset composed of 75% of the control subjects (healthy people) and 75% of the people with PD. The remaining individuals constitute the testing subset, 25% of healthy people and 25% with PD. The model parameters are determined using the training subset, and errors are computed using the testing subset. This procedure is repeated 100 times and the results are then averaged. Note that at each iteration by using the training subset, the CPCs defined in Sect. 3.1 are obtained, and then they are used to estimate the parameters of the proposed approach defined in Sects. 3.2 and 3.3. Table 1 shows the results of the classification measures of the scenarios described above.

5.2 Experimental Results

Following the previous experimental setting, the results are presented in Table 1. Approach C provides the best results in all accuracy metrics. The second best results are obtained with Approach D. Both approaches outperform approaches A and B in all accuracy metrics.

The percentage improvements provided by Approach C with respect to the remaining ones are the following. The accuracy metrics in Approach C improves the ones in Approach D only between 0.67 and 1.70%. However, it improves Approach B between 2.50 and 8.10%. The largest improvement of percentages are obtained when comparing with Approach A, leading to values between 4.76 and 15.18%.

With an illustrative purpose, Fig. 1 shows the histograms of some variables x_{ijk} that have been modeled with different number of components using mixtures of normal distributions. The bars show the relative frequency of the variables x_{ijk}, and the curves are the density of the mixture of normal distributions of w_{ik}. Note that in some variables it is clear their distribution could be modeled by a mixture of normal densities due to its multimodal nature.

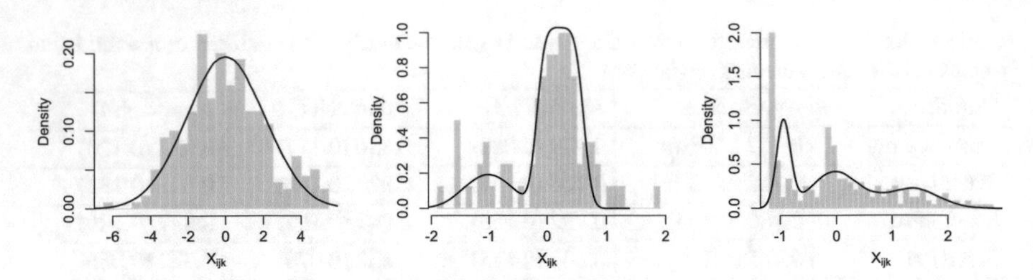

Fig. 1 Histograms and density of the mixtures of normal distributions of some variables modeled with different number of components (modes): (left) one mode for HNR group; (middle) two modes for CPC2 of $\mathcal{G}_4$; (right) three modes for PPE

6 Conclusion

A two-stage classification approach has been developed to differentiate PD subjects from healthy people by using acoustic features extracted from replicated voice recordings. Due to the high number of correlated acoustic features, CPC analysis has been considered together with a Bayesian hierarchical approach specifically designed to classify subjects by handling replicated covariates and multimodal distributions. The computational problem of generating from the posterior distribution has been avoided by applying MCMC methods and using JAGS. The results obtained improve the ones obtained in the scientific literature between 1.70 and 11.17% for accuracy rate and between 0.67 and 4.76% for AUC.

Detecting Parkinson's disease (PD) by using noninvasive low-cost tools as proposed here is a topic of interest. Developing new approaches based on the nonlinear behavior of the impaired voice production may improve the methodology based on the linear assumption. Our next research will address nonlinear classification approaches in a replication-based framework.

Acknowledgements Thanks to the anonymous participants and to Carmen Bravo and Rosa María Muñoz for carrying out the voice recordings and providing information from the people with PD. We are grateful to the *Asociación Regional de Parkinson de Extremadura* and *Confederación Española de Personas con Discapacidad Física y Orgánica* for providing support in the experiment development.

This research has been supported by UNAM-DGAPA-PAPIIT, Mexico (Project IA106416), *Ministerio de Economía, Industria y Competitividad*, Spain (Projects MTM2014-56949-C3-3-R and MTM2017-86875-C3-2-R), *Junta de Extremadura*, Spain (Projects IB16054 and GRU18108), and the *European Union* (European Regional Development Funds).

References

1. Buonaccorsi, J.P.: Measurement Error: Models. Methods and Applications. Chapman and Hall/CRC, Boca Raton, FL (2010)
2. Carroll, R.J., Roeder, K., Wasserman, L.: Flexible parametric measurement error models. Biometrics **55**(1), 44–54 (1999)

3. Carroll, R.J., Ruppert, D., Stefanski, L.A., Crainiceanu, C.M.: Measurement Error in Nonlinear Models: A Modern Perspective, 2nd edn. Chapman and Hall/CRC, Boca Raton, FL (2006)
4. Chen, M.H., Shao, Q.M., Ibrahim, J.G.: Monte Carlo Methods in Bayesian Computation. Series in Statistics. Springer, Berlin (2000)
5. Diebolt, J., Robert, C.P.: Estimation of finite mixture distributions through Bayesian sampling. J. R. Stat. Soc. Ser. B (Methodol.) **56**(2), 363–375 (1994)
6. Flury, B.: Common Principal Components and Related Models. Wiley, New York (1988)
7. Flury, B.N.: Common principal components analysis in K groups. J. Am. Stat. Assoc. **79**(388), 892–898 (1984)
8. Flury, B.N.: Theory for common principal component analysis. Ann. Stat. **14**(2), 418–430 (1986)
9. Harel, B., Cannizzaro, M., Snyder, P.J.: Variability in fundamental frequency during speech in prodromal and incipient Parkinson's disease: a longitudinal case study. Brain Cogn. **56**(1), 24–29 (2004)
10. Hariharan, M., Polat, K., Sindhu, R.: A new hybrid intelligent system for accurate detection of Parkinson's disease. Comput. Methods Programs Biomed. **113**(3), 904–913 (2014)
11. Jolliffe, I.T.: Principal Component Analysis, 2nd edn. Springer, New York (2002)
12. Little, M.A., McSharry, P.E., Hunter, E.J., Spielman, J., Ramig, L.O.: Suitability of dysphonia measurements for telemonitoring of Parkinson's disease. IEEE Trans. Biomed. Eng. **56**(4), 1015–1022 (2009)
13. Lykou, A., Ntzoufras, I.: On Bayesian LASSO variable selection and the specification of the shrinkage parameter. Stat. Comput. **23**(3), 361–390 (2013)
14. McLachlan, G.J., Krishnan, T.: The EM Algorithm and Extensions, 2nd edn. Wiley, New York (2007)
15. Naranjo, L., Pérez, C.J., Campos-Roca, Y., Martín, J.: Addressing voice recording replications for Parkinson's disease detection. Expert Syst. Appl. **46**, 286–292 (2016)
16. Naranjo, L., Pérez, C.J., Martín, J., Campos-Roca, Y.: A two-stage variable selection and classification approach for Parkinson's disease detection by using voice recording replications. Comput. Methods Programs Biomed. **142**, 147–156 (2017)
17. O'Hara, R.B., Sillanpää, M.J.: A review of Bayesian variable selection methods: what, how and which. Bayesian Anal. **4**(1), 85–118 (2009)
18. Orozco-Arroyave, J.R., Arias-Londoño, J.D., Vargas-Bonilla, J.F., Nöth, E.: Analysis of speech from people with Parkinson's disease through nonlinear dynamics. In: Drugman, T., Dutoit, T. (eds.) Advances in Nonlinear Speech Processing. Lecture Notes in Artificial Intelligence, vol. LNAI 7911, pp. 112–119. Springer, Berlin (2013)
19. Pahuja, G., Nagabhushan, T.N.: A comparative study of existing machine learning approaches for Parkinson's disease detection. IETE J. Res. 1–11 (2018)
20. Park, T., Casella, G.: The Bayesian LASSO. J. Am. Stat. Assoc. **103**(482), 681–686 (2008)
21. Pérez, C.J., Naranjo, L., Martín, J., Campos-Roca, Y.: A latent variable-based Bayesian regression to address recording replication in Parkinson's disease. In: EURASIP: Proceedings of the 22nd European Signal Processing Conference (EUSIPCO-2014), pp. 1447–1451. IEEE, Lisbon, Portugal (2014)
22. Plummer, M.: Penalized loss functions for Bayesian model comparison. Biostatistics **9**(3), 523–539 (2008)
23. Richardson, S., Green, P.J.: On Bayesian analysis of mixtures with an unknown number of components. J. R. Stat. Soc. Ser B (Methodol.) **59**(4), 731–792 (1997)
24. Rockova, V., Lesaffre, E., Luime, J., Löwenberg, B.: Hierarchical Bayesian formulations for selecting variables in regression models. Stat. Med. **31**, 1221–1237 (2012)
25. Roeder, K., Wasserman, L.: Practical Bayesian density estimation using mixtures of normals. J. Am. Stat. Assoc. **92**(439), 894–902 (1997)
26. Saifer, A., Ali, D.S.M.: A review on Parkinson's disease diagnosis through speech. Int. J. Sci. Res. Sci. Technol. **4**(5), 36–45 (2018)
27. Smith, B.J.: BOA: an R package for MCMC output convergence assessment and posterior inference. J. Stat. Softw. **21**(11), 1–37 (2007)

28. Spiegelhalter, D., Best, N., Carlin, B., van der Linde, A.: Bayesian measures of model complexity and fit. J. R. Stat. Soc. Ser. B **64**, 583–639 (2002)
29. Tsanas, A., Little, M.A., McSharry, P.E., Spielman, J., Ramig, L.O.: Novel speech signal processing algorithms for high-accuracy classification of Parkinson's disease. IEEE Trans. Biomed. Eng. **59**(5), 1264–1271 (2012)
30. Tysnes, O.B., Storstein, A.: Epidemiology of Parkinson's disease. J. Neural Transm. **124**(8), 901–905 (2017)

Calibration of Population Growth Mathematical Models by Using Time Series

Francisco Novoa-Muñoz, Sergio Contreras Espinoza, Aníbal Coronel Pérez and Ian Hess Duque

Abstract In this paper, we study the problem of coefficients identification in population growth models. We consider that the dynamics of the population is described by a system of ordinary differential equations of susceptible-infective-recovered (SIR) type, and we assume that we have a discrete observation of infective population. We construct a continuous observation by applying time series and an appropriate fitting to the discrete observation data. The identification problem consists in the determination of different parameters in the governing equations such that the infective population obtained as solution of the SIR system is as close as to the observation. We introduce a reformulation of the calibration problem as an optimization problem where the objective function and the restriction are given by the comparison in the L_2-norm of theoretical solution of the mathematical model and the observation, and the SIR system governing the phenomenon, respectively. We solve numerically the optimization problem by applying the gradient method where the gradient of the cost function is obtained by introducing an adjoint state. In addition, we consider a numerical example to illustrate the application of the proposed calibration method.

Keywords Calibration · Inverse problems · Time series · SIR models

---

F. Novoa-Muñoz (✉) · S. C. Espinoza
GMA, Departamento de Estadística, Facultad de Ciencias, Universidad del Bío-Bío, Concepción, Chile
e-mail: fnovoa@ubiobio.cl

S. C. Espinoza
e-mail: scontre@ubiobio.cl

A. C. Pérez · I. H. Duque
GMA, Departamento de Ciencias Básicas, Facultad de Ciencias, Universidad del Bío-Bío, Chillán, Chile
e-mail: acoronel@ubiobio.cl

I. H. Duque
e-mail: ihess@egresados.ubiobio.cl

I. Antoniano-Villalobos et al. (eds.), *Selected Contributions on Statistics and Data Science in Latin America*, Springer Proceedings in Mathematics & Statistics 301,
https://doi.org/10.1007/978-3-030-31551-1_8

1 Introduction

Throughout the human history, there are diseases that due to their characteristics suddenly affect a large part of the population of a certain region, generating considerable morbidity and mortality [19]. This type of disease is called epidemic and is developed in populations that acquire a certain population density [6]. However, we observe that the term epidemic is also used sometimes in the case of noninfectious diseases that are of population scale, for instance, the dynamic of diabetes. In this work, we will analyze the epidemics due to infectious diseases and within them those that are not transmitted through vectors.

From the historic point of view, we notice that the epidemics caused several health problems on a population scale, jeopardizing the survival of different civilizations. For instance, the bubonic plague [25] and its famous plagues developed in the old Egypt [24], the epidemic of plague in Athens, Typhoid and Syracuse [4] in the old Greece, the black plague that affected the whole of Europe [34], the epidemics due to the meeting between Europeans and Native Americans in America [26], and the outbreaks of cholera due to the contamination of water with fecal matter [26]. We remark that the consequences of the plagues in America are even more important than those occurring in Europe [15].

In the last decades, there are several efforts to understand the dynamics of diseases caused by epidemic. Nowadays, the mathematical epidemiology is one of the most important branches of bio-mathematics [11, 13]. Moreover, we observe that there are several kinds of mathematical models. For instance, there are mathematical models in terms of discrete mathematics, deterministic or even stochastic ordinary differential systems or partial differential equations, and statistical theory.

The mathematical modeling of the population dynamics for infectious diseases is a standard or classical problem in differential equations theory [21, 27]. We observe that the earliest published paper on mathematical modeling of spread of disease was carried out in 1766 by Daniel Bernoulli. Trained as a physician, Bernoulli created a mathematical model to defend the practice of inoculating against smallpox [18]. According to Pesco [26], in 1927 Kermack and Mc Kendrick published an article in which they proposed a mathematical model, implemented in differential equations, that simulates the transmission of an infectious disease. This model divides the population into compartments according to the epidemiological status of the individuals, classifying them as susceptible (S), infected (I), and recovered (R), which is currently known as the SIR model.

On the other hand, related with statistical methods, we notice that inferential methods have also been developed to evaluate the correlation between epidemiological data and possible indicators of risk or health policies [1, 23]. Nowadays, epidemiology is used to describe the clinical spectrum of a disease, to know the transmission mechanisms of the disease, to know the natural history of biological systems, to make population diagnoses, to identify factors that produce the risk of acquiring the disease, and to test the effectiveness of different mitigation strategies. It is mainly used to generate information required by public health professionals to develop, implement, and evaluate health policies [12]. However, in the best of our knowledge, there are no works related with time series theory applied to model calibration in epidemiology. An exception and an advance in this research line is the recent work [9].

In the processes of mathematical modeling by ordinary differential equations, there are at least four phases: abstraction, simplification or mathematical model formulation, solution or analysis, and validation [14]. Particularly, the phase of validation requires the solution of problem well known as the mathematical model calibration.

In a broad sense, parameter calibration means that we want to find (or to calculate) some unknown constants or functions (called model parameters) from some given observations for the solution model. The mathematical concept of calibration or identification is equivalent to that of estimation in statistics. In practice, these problems can be solved by applying the inverse problem methodologies [14]. We remark that, although the estimation (calibration or identification) of unknown parameters has a significant practical importance, there are several problems which are not enough investigated due, for instance, to the lack of results on the uniqueness of the solution of the inverse problem, i.e., while the direct problem may have a unique solution, the inverse problem does not usually have the same property [21, 30]. Moreover, we observe that the inverse problem is crucial for calibrating the model and for controlling the model parameters. Approaches involving inverse problems can be successfully applied to a variety of important biological processes, including the spread of infectious diseases, allowing epidemiologists and public health specialists to predict the time interval between epidemics [5, 21].

The aim of this paper is the identification of certain coefficients (or parameters) in the ordinary differential equations system of SIR type investigated by Bai and Zhou [3] by using the inverse problem methodologies and the time series theory. We start by defining a continuous observation using the time series and an interpolation of discrete data. Then, we define an optimization problem for an appropriate cost function which is equivalent to the inverse problem. To solve the minimization problem, we apply the gradient method where the gradient of the cost function is calculated by the introduction of an adjoint state.

Among some previous and related works with the topic of parameter identification in epidemiological models, we can refer to [16–18, 20, 22, 29, 33]. The models considered by the authors are systems of stochastic differential equations and the notion of parameters adopted by them is given by the context of statistics theory. Thus, the methodologies are not comparable with the ones presented in this paper, since the model is a deterministic model and the notion of parameter is used to define the coefficients of the system.

The rest of the paper is organized as follows. In Sect. 2, we present the notation and precise definition of the direct problem. In Sect. 3, we define the inverse problem. In Sect. 4, we present the inverse problem solution methodology. In Sect. 5, we present a numerical experiment. Finally, in Sect. 6, we summarize some conclusions.

2 The Direct Problem

Let $S(t)$ be the number of susceptible individuals, $I(t)$ be the number of infective individuals, and $R(t)$ be the number of recovered individuals at time $t \in [0, T]$, respectively. According to the writing by Bai and Zou [3], "After studying the cholera epidemic spread in Bari in 1973, Capasso and Serio introduced the saturated incidence rate $\beta SI(1+kI)^{-1}$ into epidemic model," where βI measures the infection force of the disease and $(1+kI)^{-1}$ with $k > 0$ describes the psychological effect or inhibition effect from the behavioral change of the susceptible individuals with the increase of the infective individuals. This incidence rate seems more reasonable than the bilinear incidence rate βSI, because it includes the behavioral change and crowding effect of the infective individuals and prevents the unboundedness of the contact rate.

The treatment is an important way to reduce the disease spread, such as measles, tuberculosis, and flu [32]. In classical epidemic models, the treatment rate of infectives is assumed to be proportional to the number of the infectives. The proportional assumption will lead to very fast increase of the treatment resource. In fact, every community has a suitable capacity for treatment. If it is too large, the community pays for unnecessary cost. If it is too small, the community has a higher risk of disease outbreak. It is realistic to maintain a suitable capacity of disease treatment. Wang and Ruan [31] introduced a treatment function $h(I)$, which is a positive constant m for $I > 0$, and zero for $I = 0$. This seems more reasonable when we consider the limitation of the treatment resource of a community.

Bai and Zhou [3] formulated a nonautonomous SIR epidemic model with saturated incidence rate and constant removal rate by introducing the periodic transmission rate $\beta(t)$. The general model is formulated as follows:

$$\left.\begin{aligned} \frac{d}{dt}S(t) &= \Lambda - \mu\, S(t) - \frac{\beta(t)S(t)\,I(t)}{1+kI(t)}, \\ \frac{d}{dt}I(t) &= \frac{\beta(t)S(t)\,I(t)}{1+kI(t)} - (\mu+\gamma)\,I(t) - h(I(t)), \\ \frac{d}{dt}R(t) &= \gamma I(t) + h(I(t)) - \mu\, R(t). \end{aligned}\right\} \tag{1}$$

Here, Λ is the recruitment rate, μ is the natural death rate, γ is the recovery rate of the infective population, and $\beta(t)$ is the transmission rate at time t. Now, noticing that the first two equations in (1) are independent of the third one, and the dynamic behavior of (1) is trivial when $I(t_0) = 0$ for some $t_0 > 0$, Bai and Zhou [3] considered only the first two equations with $I > 0$. Thus, these researchers restricted their study to the model given by

$$\left.\begin{aligned} \frac{d}{dt}S(t) &= \Lambda - \mu\, S(t) - \frac{\beta(t)S(t)\,I(t)}{1+kI(t)}, \\ \frac{d}{dt}I(t) &= \frac{\beta(t)S(t)\,I(t)}{1+kI(t)} - (\mu+\gamma)\,I(t) - m. \end{aligned}\right\} \tag{2}$$

In the terminology of inverse problems, we have that the direct problem is given by system (2) with some appropriate initial conditions for S and I. More precisely

Definition 1 The direct problem is formulated as follows: given the constants $T, \Lambda, \mu, k, m, S_0, I_0$ and the function β, find the functions S and I satisfying the system (2) on the interval $]0, T]$ and the initial condition $(S, I)(0) = (S_0, I_0)$.

The direct problem is well-posed since it is the standard Cauchy problem for an ordinary differential system where the right-hand side is a locally Lipschitz function.

3 The Inverse Problem

The inverse problems consist in the determination of μ and γ in the system (2) from a distribution of the number of infected individuals I^{obs} and such that the infected solution of the direct problem for μ and γ, denoted as $I_{\mu,\gamma}$, is "as close as" to I^{obs}. The term "as close as" is numerically precise by considering the L_2-norm of the distance of $I_{\mu,\gamma}$ and I^{obs}. However, we observe that I^{obs} is not defined on the whole time interval. Then, to extend I^{obs} continuously we apply time series. Then we precise the definition of the optimization problem.

To precise the application of time series, we consider the numeric values for the parameters used by Bai and Zhou [3] to investigate the stability of the periodic solution of (2) with given parameter values and small degree seasonal fluctuation in transmission rate. We set that $\Lambda = 400$, $k = 0.01$, $\mu = 0.02$, $\gamma = 0.04$, $m = 10$, and $\beta(t) = 0.00006 + \varepsilon \sin(\pi t/3)$, where $0 \leq \varepsilon < 0.00006$. Then, the system (2) becomes

$$\left.\begin{aligned} \frac{d}{dt}S(t) &= 400 - 0.02\,S(t) - \frac{3\,[0.2 + \varepsilon \sin(\pi t/3)]\,S(t)\,I(t)}{10000 + 100\,I(t)}, \\ \frac{d}{dt}I(t) &= \frac{3\,[0.2 + \varepsilon \sin(\pi t/3)]\,S(t)\,I(t)}{10000 + 100\,I(t)} - (0.02 + 0.04)\,I(t) - 10, \\ S(0) &= 14000, \quad I(0) = 600. \end{aligned}\right\} \tag{3}$$

Now, we solve numerically system (3) and obtain the discrete synthetic observed data which is shown on Fig. 1. The next step is to adjust this data by a continuous function. Indeed, due to the structure of the data, we deduced that it would be very useful to use time series to find the best model that fits such data. Figure 2 shows the graph of the time series associated with the data. To be more precise, as the series shown in Fig. 2 shows a lot of variability, the first thing we did was to apply a transformation to the data. Then, using the programming language R [28] and the tseries library next to the Arima command, we obtained a two-differentiated $AR(1)$ model [7], in which results are summarized in the following expression:

$$(1 - B)^2(1 - B^6)(1 - 0.512B)Y_t = \varepsilon_t, \quad \varepsilon_t \sim N(0, \sigma^2),$$

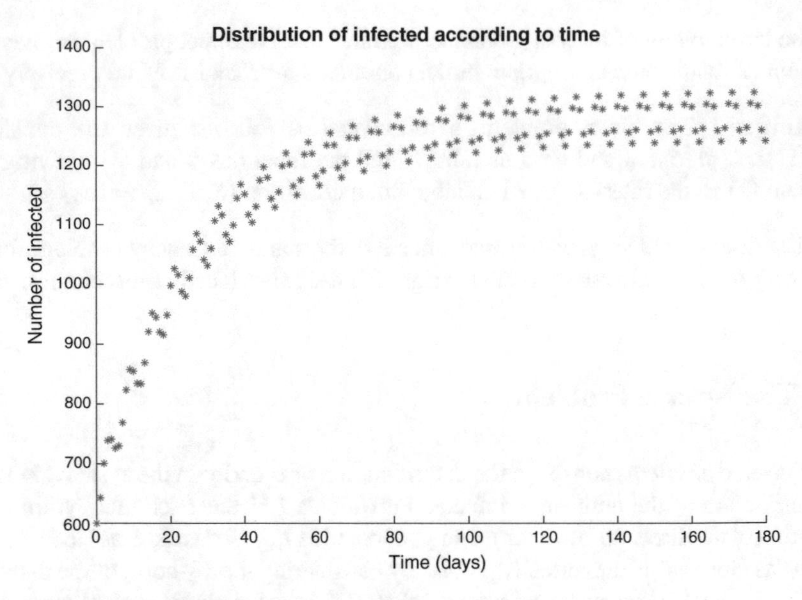

Fig. 1 Plot of the number of infected population solution of (3)

where B is the lag operator. Remembering that $BY_t = Y_{t-1}$, it turns out

$$Y_t = 2.512Y_{t-1} - 2.024Y_{t-2} + 0.512Y_{t-3} + Y_{t-6} \\ -2.512Y_{t-7} + 2.024Y_{t-8} - 0.512Y_{t-9} + \varepsilon_t, \quad (4)$$

where ε_t is a white noise $N(0, \sigma^2)$.

Then, using this time series and an appropriate interpolation we construct the function $I^{obs}(t)$ on $[0, T]$.

We observe that model (4) corresponds to the synthetic data obtained by simulation of (3). Then for other particular cases of system (2), we proceed analogously to construct the corresponding time series model and the appropriate I^{obs} continuous function.

We reformulate the inverse problem like an optimal control problem. The optimization problem is now formulated as follows: the objective function J depending on the variables μ and γ is the least squares cost function and the restriction is the initial value problem for the system (2) with some parameters μ and γ. More precisely we have the following definition.

Definition 2 The inverse problem is defined by the optimization problem:

$$\text{Minimize } J(\mu, \gamma) = \delta \left\| I_{\mu,\gamma} - I^{obs} \right\|^2_{L_2(0,T)} := \delta \int_0^T (I_{\mu,\gamma} - I^{obs})^2(t)dt \quad (5)$$

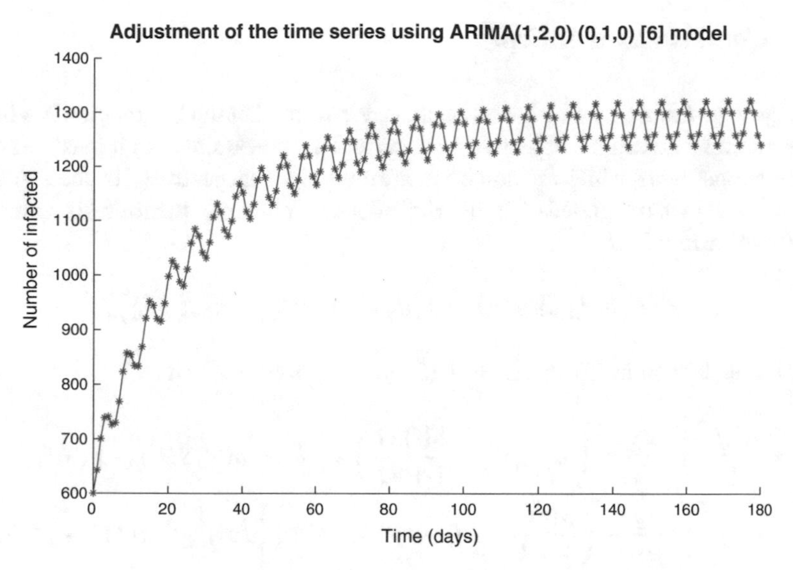

Fig. 2 Plot of the adjustment of infected population solution of (3) by the time series to an ARIMA$(1, 2, 0) \times (0, 1, 0)_6$ model

subject to $I_{\mu,\gamma}$ solution of system (2) a given initial condition $(S, I)(0) = (S_0, I_0)$, I^{obs} is the continuous observation constructed with a time series and an appropriate interpolation, and δ a positive constant.

We remark two facts. First, the objective function J measures the distance between two functions, and therefore it must be minimized and is called cost. One of the functions involved in this cost is that which solves the initial value problem and the other is the one that adjusts the data through the series of time already found. Second, we observe that the existence of solutions for the inverse problem can be derived by applying the continuous dependence of (S, I) with respect to (μ, γ). However, the uniqueness of the inverse problem solution or the proof of a unique global optimizer for J is difficult to get.

4 Methodology of the Solution for the Inverse Problem

In order to solve the optimization problem, we apply the gradient method where the gradient of the cost function is obtained by the introduction of an adjoint state. First, we deduce the gradient of the cost function when all data are continuous and then we mimic the process to introduce a discrete gradient.

4.1 Continuous Gradient

We apply the Lagrange multipliers method. We recall that the Lagrangian is a linear combination between the objective function and the constraints. As it must be scalar then the constraints, which are functions, are multiplied, respectively, by the functions p and q, and then integrated over the whole domain, obtaining the following function to be minimized:

$$\mathscr{L}\big(S_{\mu,\gamma}, I_{\mu,\gamma}; p, q\big) = J(\mu,\gamma) - E\big(S_{\mu,\gamma}, I_{\mu,\gamma}; p, q\big),$$

where J is defined in (5) and $E := E(S_{\mu,\gamma}, I_{\mu,\gamma}; p, q)$ is given by

$$\begin{aligned} E = &- \int_0^T \left[S\frac{dp}{dt} + \left(\Lambda - \mu S - \frac{\beta(t)SI}{1+kI} \right) p \right] d\tau + p(T)S(T) - p(0)S_0 \\ &- \int_0^T \left[I\frac{dq}{dt} + \left(\frac{\beta(t)SI}{1+kI} - (\mu+\gamma)I - m \right) q \right] d\tau + q(T)I(T) - q(0)I_0. \end{aligned}$$

Thus, we have a classic minimization problem for $\mathscr{L}$, and therefore we can apply the first-order optimal conditions to deduce the adjoint state, i.e., we calculate the gradient of the Lagrangian with respect to the variables μ and γ, and select the states (p, q) such that the gradient of $\mathscr{L}$ vanishes. We note that

$$\begin{aligned} \frac{d\mathscr{L}}{d\mu} &= \frac{\partial \mathscr{L}}{\partial S}\frac{\partial S}{\partial \mu} + \frac{\partial \mathscr{L}}{\partial I}\frac{\partial I}{\partial \mu} + \frac{\partial \mathscr{L}}{\partial \mu} \equiv 0, \\ \frac{d\mathscr{L}}{d\gamma} &= \frac{\partial \mathscr{L}}{\partial S}\frac{\partial S}{\partial \gamma} + \frac{\partial \mathscr{L}}{\partial I}\frac{\partial I}{\partial \gamma} + \frac{\partial \mathscr{L}}{\partial \gamma} \equiv 0. \end{aligned}$$

The calculus of $\partial_\mu S$, $\partial_\mu I$, $\partial_\gamma S$, and $\partial_\gamma I$ is difficult to develop directly since the functions S and I do not depend explicitly on μ and γ and the strategy is to select p and q such that $\partial_S \mathscr{L} = \partial_I \mathscr{L} = 0$. We observe that

$$\begin{aligned} \frac{\partial \mathscr{L}}{\partial S} &= -\frac{\partial E}{\partial S} = \int_0^T \left[\frac{dp}{dt} - \left(\mu + \frac{\beta(t)I}{1+kI} \right) p + \frac{\beta(t)Iq}{1+kI} \right] d\tau - p(T) \\ \frac{\partial \mathscr{L}}{\partial I} &= \frac{\partial J}{\partial I} - \frac{\partial E}{\partial I} \\ &= 2\delta \int_0^T (I - I^{obs}) + \frac{dq}{dt} - \left[\frac{\beta(t)S}{(1+kI)^2} p + \frac{\beta(t)S}{(1+kI)^2} q + (\mu+\gamma)q \right] d\tau - q(T). \end{aligned}$$

Thus, a necessary condition for p and q such that $\partial_S \mathscr{L} = \partial_I \mathscr{L} = 0$ is given by

$$\left.\begin{aligned} \frac{dp}{dt} &= \mu\, p + \frac{\beta(t)\, I}{1+kI}(p-q), \\ \frac{dq}{dt} &= (\mu+\gamma)\, q + \frac{\beta(t)\, S}{(1+kI)^2}(p-q) + 2\delta(I^{obs} - I), \\ p(T) &= q(T) = 0. \end{aligned}\right\} \tag{6}$$

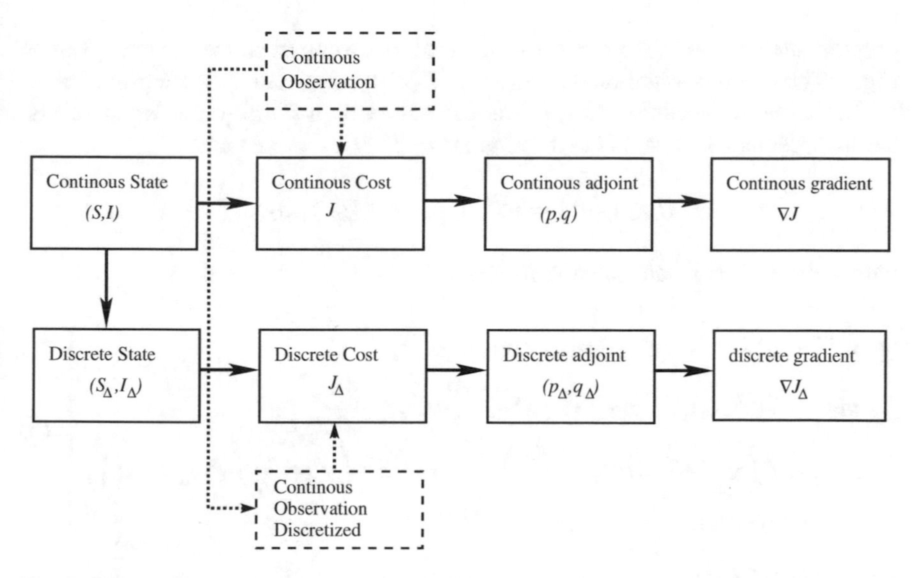

Fig. 3 Scheme of the gradient calculus methodology. The definition of (S, I), J, (p, q), and ∇J is given in (2), (5), (6), and (7), respectively. Now, the definition of (S_Δ, I_Δ), J_Δ, (p_Δ, q_Δ), and ∇J_Δ is presented in (8), (9), (11), and (12), respectively

The backward problem (6) is called the continuous adjoint state. It gives rise to the so-called continuous gradient, which arises equal to zero the derivatives of the Lagrangian, obtaining

$$\nabla J = \nabla E = -\int_0^T (Sp, Iq)d\tau. \tag{7}$$

The gradient given in (7) can be used to solve numerically the optimization problem. However, there are several suggestions to use a discrete gradient obtained by a similar methodology instead of direct discretization for ∇J in (7), see, for instance, [8, 10].

4.2 *Discrete Gradient*

The discretization of (7) typically develops numerical instabilities [8, 10]. Then, the strategy is to obtain a discrete gradient by a similar methodology to that applied to get (7). We recall that the three steps are the following: (i) construct the continuous cost function given on (5), (ii) apply a Lagrangian formulation to define the continuous adjoint state given on (6), and (iii) use the continuous and the adjoint state to define the continuous gradient given on (7), see the upper part on the scheme given on Fig. 3. Then, to obtain the discrete we discretize the continuous state variables and proceed analogously to the steps (i)–(iii), see the lower part on the scheme given on Fig. 3. The specific definitions of (S_Δ, I_Δ), J_Δ, (p_Δ, q_Δ), and ∇J_Δ are given below.

The numerical solution (S_Δ, I_Δ) is calculated by a fourth-order Runge–Kutta method. Let us select $N \in \mathbb{N}$ and define $\Delta t = T/N$, $t_n = n\Delta t$ and

$$(S_\Delta, I_\Delta)(t) = (S^n, I^n), \quad t \in [t_n, t_{n+1}[,$$

with $\mathbf{x}^n := (S^n, I^n)^t$ calculated as follows:

$$\left.\begin{aligned}
\mathbf{x}^{n+1} &= \mathbf{x}^n + \frac{\Delta t}{6}\Big(\mathbf{m}_1 + 2\mathbf{m}_2 + 2\mathbf{m}_3 + \mathbf{m}_4\Big),\\
\mathbf{m}_1 &= \mathbf{f}(\mathbf{x}^n, t_n), \qquad \mathbf{m}_2 = \mathbf{f}\left(\mathbf{x}^n + \frac{\Delta t}{2}\mathbf{m}^1, t_n + \frac{\Delta t}{2}\right),\\
\mathbf{m}_3 &= \mathbf{f}\left(\mathbf{x}^n + \frac{\Delta t}{2}\mathbf{m}^2, t_n + \frac{\Delta t}{2}\right), \qquad \mathbf{m}_4 = \mathbf{f}\left(\mathbf{x}^n + \frac{\Delta t}{2}\mathbf{m}^3, t_n + \Delta t\right),\\
\mathbf{x}^0 &= (S_0, I_0),
\end{aligned}\right\} \tag{8}$$

where $\mathbf{f}$ is defined by

$$\mathbf{f}\begin{pmatrix} S \\ I \\ t \end{pmatrix} = \begin{pmatrix} \Lambda - \mu\, S - \dfrac{\beta(t) S\, I}{1 + kI} \\ \dfrac{\beta(t) S\, I}{1 + kI} - (\mu + \gamma)\, I - m \end{pmatrix},$$

i.e., $\mathbf{f}$ is the right side of the system (2). Thus, (8) with $\mathbf{x}^n := (S^n, I^n)^t$ is the discretization of (2), which is called the discrete state.

The discrete cost function J_Δ is given by

$$J_\Delta(\mu, \gamma) = \delta \Delta t \sum_{n=0}^{N} \left(I^n - I^{obs,n}\right)^2, \tag{9}$$

where I^n is obtained by the numerical scheme (8) and $I^{obs,n}$ is the evaluation of the continuous observation on the nodes of the mesh, i.e., $I^{obs,n} = I^{obs}(t_n)$. Note that J_Δ given on (9) is the natural discretization of J given on (5). Thus, we have that

$$\text{Minimize } \; J_\Delta(\mu, \gamma) \; \text{ subject to } I_\Delta \text{ solution of the numerical scheme (2)} \tag{10}$$

is the discrete version or the discretization of the optimization problem (5).

In order to define the discrete adjoint state (p_Δ, q_Δ), we apply the Lagrange multipliers method. We define the discrete Lagrangian

$$\mathscr{L}_\Delta(S_\Delta, I_\Delta; p_\Delta, q_\Delta) = J_\Delta(\mu, \gamma) - E_\Delta(S_\Delta, I_\Delta; p_\Delta, q_\Delta),$$

where J_Δ is defined in (9) and $E_\Delta := E(S_\Delta, I_\Delta; p_\Delta, q_\Delta)$ is given by

$$E_{\Delta} = \sum_{n=0}^{N-1} \left[\mathbf{x}^n \left(\mathbf{p}^{n-1} - \mathbf{p}^n \right) - \frac{\Delta t}{6} \left(\mathbf{m}_1 + 2\mathbf{m}_2 + 2\mathbf{m}_3 + \mathbf{m}_4 \right) \right] + \mathbf{x}^N \mathbf{p}^{N-1} - \mathbf{x}^0 \mathbf{p}^{-1}.$$

Thus, by analogous arguments to the continuous case, we need to determine (p_{Δ}, q_{Δ}) such that $\partial_{S^n} \mathscr{L}_{\Delta} = \partial_{I^n} \mathscr{L}_{\Delta} = 0$. Thus, from differentiation of $\mathscr{L}_{\Delta}$ we obtain that $\mathbf{p}^n := (p^n, q^n)^t$ can calculated by the scheme

$$\left. \begin{aligned} \mathbf{p}^{n-1} &= \mathbf{p}^n + \frac{\Delta t}{6} \left(\mathbf{n}_1 + 2\mathbf{n}_2 + 2\mathbf{n}_3 + \mathbf{n}_4 \right) + 2\delta \Delta t \mathbf{g}(\mathbf{x}^n), \\ \mathbf{n}_1 &= \partial \mathbf{f}(\mathbf{x}^n, t_n), \qquad \mathbf{n}_2 = \partial \mathbf{f} \left(\mathbf{x}^n + \frac{\Delta t}{2} \mathbf{m}^1, t_n + \frac{\Delta t}{2} \right), \\ \mathbf{n}_3 &= \partial \mathbf{f} \left(\mathbf{x}^n + \frac{\Delta t}{2} \mathbf{m}^2, t_n + \frac{\Delta t}{2} \right), \qquad \mathbf{n}_4 = \partial \mathbf{f} \left(\mathbf{x}^n + \frac{\Delta t}{2} \mathbf{m}^3, t_n + \Delta t \right), \\ \mathbf{p}^N &= (0, 0), \end{aligned} \right\} \tag{11}$$

where

$$\partial \mathbf{f} \begin{pmatrix} S \\ I \\ t \end{pmatrix} = \begin{pmatrix} -\mu - \dfrac{\beta(t)\, I}{1 + kI} \\ \dfrac{\beta(t) S}{(1 + kI)^2} - (\mu + \gamma) \end{pmatrix}, \qquad \mathbf{g} \begin{pmatrix} S \\ I \end{pmatrix} = \begin{pmatrix} 0 \\ I - I^{obs,n} \end{pmatrix},$$

and $\mathbf{x}^n$ is calculated by (8). The scheme (11) is called the discrete adjoint state.

The discrete gradient ∇J_{Δ} is calculated by

$$\nabla J_{\Delta} = \frac{\Delta t}{6} \sum_{n=0}^{N-1} \left[\nabla_{\mu,\gamma} \mathbf{m}_1 + 2\nabla_{\mu,\gamma} \mathbf{m}_2 + 2\nabla_{\mu,\gamma} \mathbf{m}_3 + \nabla_{\mu,\gamma} \mathbf{m}_4 \right], \tag{12}$$

where $\mathbf{m}_i$ for $i = 1, \ldots, 4$ are defined in (8). The gradient given in (12) is used to solve numerically the inverse problem.

5 Numerical Results

In this section, we present a numerical result for estimating the value of parameters μ and γ from synthetic observation data. We consider the system (3) and by a numerical simulation we obtain a discrete observation. Then, by the process indicated on Sect. 3 we construct I^{obs} on $[0, T]$, see also Figs. 1 and 2. To be more precise, after processing the observation by a time series technique, we fit the discrete observation data by $I^{obs} : [0, T] \in \mathbb{R}^+$ defined by

$$\begin{aligned} I^{obs}(t) = a_1 \sin(b_1 t + c_1) + a_2 \sin(b_2 t + c_2) + a_3 \sin(b_3 t + c_3) \\ + a_4 \sin(b_4 t + c_4) + a_5 \sin(b_5 t + c_5), \end{aligned} \tag{13}$$

where the values of a_i, b_i, and c_i are given by

$$a_1 = 2412, \quad a_2 = 1457, \quad a_3 = 430.7, \quad a_4 = 114.3, \quad a_5 = 40.3,$$
$$b_1 = 0.01641, \quad b_2 = 0.02814, \quad b_3 = 0.04432, \quad b_4 = 0.05374, \quad b_5 = 1.047,$$
$$c_1 = -0.3434, \quad c_2 = 1.491, \quad c_3 = 2.903, \quad c_4 = 5.065, \quad c_5 = -1.513.$$

The graph of I^{obs} given in (13) is the curve labeled as real parameters in Figs. 4 and 5.

For identification, we use the gradient method where the gradient of the cost function is defined by (12). To be more precise, we proceed to the identification with the gradient method using the curvature information [2]:

$$\mathbf{e}^{k+1} = \mathbf{e}^k - \lambda_k \nabla J(\mathbf{e}^k), \quad \text{with} \quad \mathbf{e}^k = (\mu_k, \lambda_k),$$
$$\lambda_k = \frac{\|\nabla J_\Delta(\mathbf{e}^k)\|\hat{\varepsilon}^2}{|J_\Delta(\mathbf{e}^k - \hat{\varepsilon}\nabla J_\Delta(\mathbf{e}^k)) - 2J_\Delta(\mathbf{e}^k) + J_\Delta(\mathbf{e}^k + \hat{\varepsilon}\nabla J_\Delta(\mathbf{e}^k))|}.$$

Here J_Δ is calculated by (9). Moreover, we remark that in our numerical simulations we consider that $\hat{\varepsilon} = 10E-6$ in the definition of λ_k and $\delta = 10E-12$ in the definition of J_Δ and the discrete adjoint state. The numerical value of the parameters is given in Table 1. The infected curve for the initial guess parameters is labeled as initial estimate of the parameters as shown in Fig. 4. The infected curve for the identified parameters is labeled as final estimate of the parameters shown in Fig. 5.

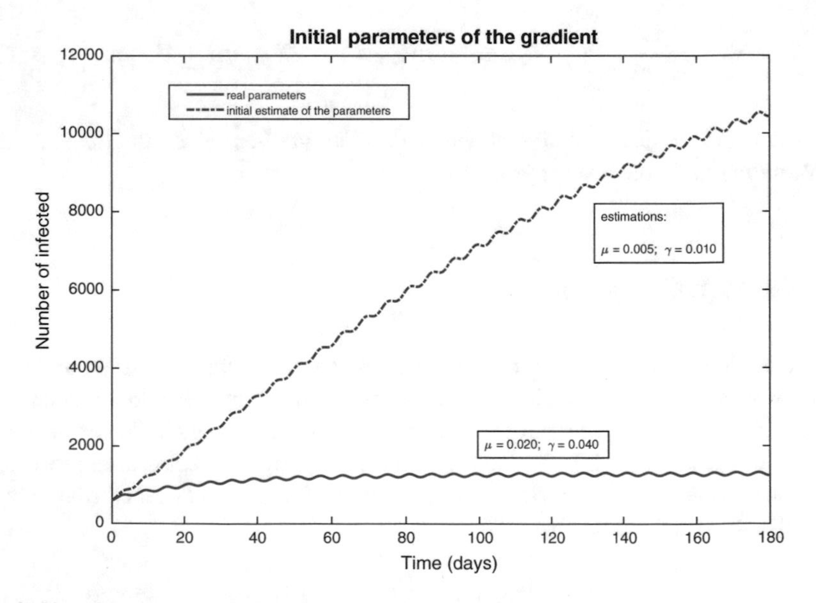

Fig. 4 Plot of the number of infected people using the real parameters and an initial estimate

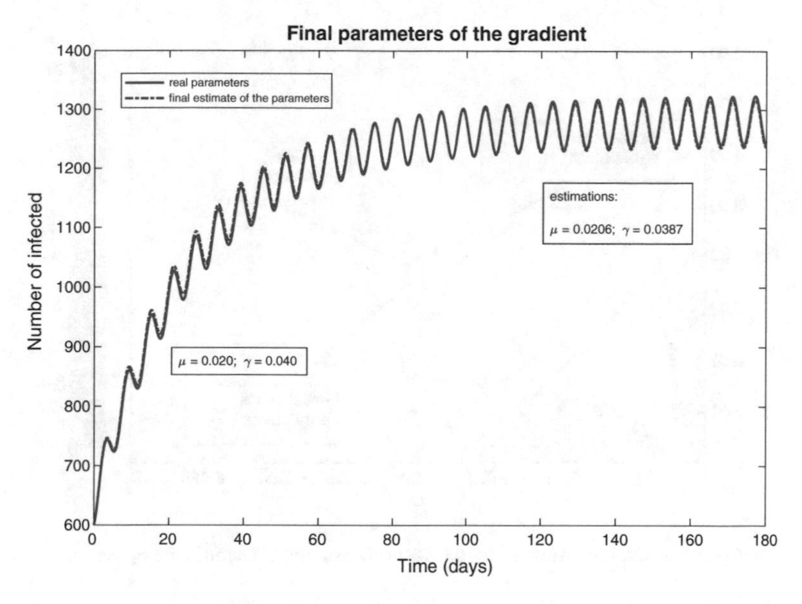

Fig. 5 Plot of the number of infected using the real parameters and a final estimate

Table 1 Numerical value of the parameters

	μ	γ
Observation parameters	0.0200	0.0400
Initial guess parameters	0.0005	0.0100
Identified parameters	0.0206	0.0387

Finally, Fig. 6 illustrates the path followed by the gradient method in order to descend the values in the cost function until reaching the point where the optimum is achieved.

6 Conclusions

In this paper, we have introduced a methodology, based on discrete gradient method and time series, for parameter identification or model calibration in ordinary differential equation systems. Although the content of this research focuses on a specific ordinary differential equations system, we can deduce that the proposed method can be generalized for identification of coefficients in other types of system. Moreover, in this study, we have applied the numerical identification for synthetic observation data and expect to apply the methodology in the model calibration when the experimental data is obtained in laboratory experiments.

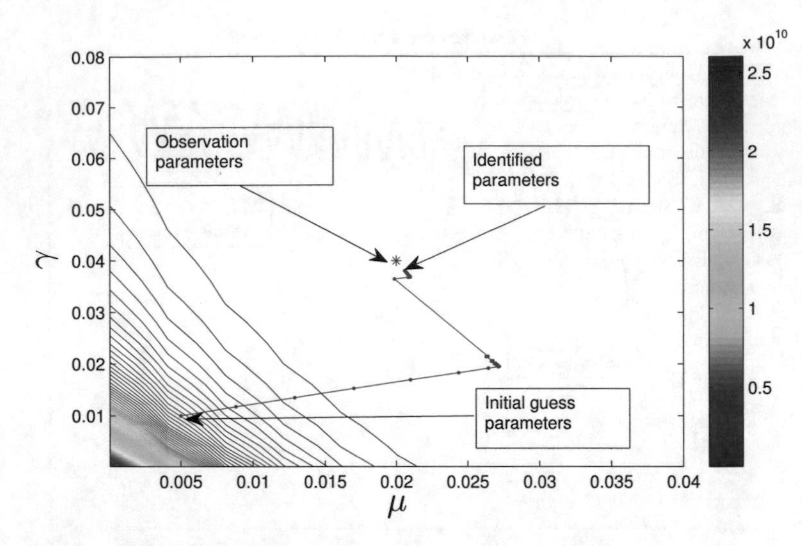

Fig. 6 Plot graph of the path traveled by the cost function until reaching the optimum

Acknowledgements We thank to research projects DIUBB 172409 GI/C and FAPEI at U. del Bío-Bío, Chile. AC thanks to the research project DIUBB 183309 4/R at U. del Bío-Bío, Chile. IH thanks to the program "Becas de doctorado" of Conicyt-Chile. Ian Hess and Francisco Novoa-Muñoz would thank the support of the program "Fortalecimiento del postgrado" of the project "Instalación del Plan Plurianual UBB 2016–2020"

References

1. Akmatov, M.K., Kretzschmar, M., Krämer, A., Mikolajczyk, R.T.: Timeliness of vaccination and its effects on fraction of vaccinated population. Vaccine **26**(31), 3805–3811 (2008)
2. Alvarez, F., Cabot, M.: Steepest descent with curvature dynamical system. J. Optim. Theory Appl. **120**, 247–273 (2004). https://doi.org/10.1023/B:JOTA.0000015684.50827.49
3. Bai, Z., Zhou, Y.: Existence of two periodic solutions for a non-autonomous SIR epidemic model. Appl. Math. Model. **35**, 382–391 (2011). https://doi.org/10.1016/j.apm.2010.07.002
4. Barriga, G., Hernández, E.: Aspectos actuales de las infecciones emergentes y reemergentes. Revista Latinoamericana de Patología Clínica y Medicina de Laboratorio **62**(3), 174–182 (2015)
5. Bauch, C., Earn, D.: Interepidemic intervals in forced and unforced SEIR models. In: Ruan, S., Wolkowicz, G.S., Wu, J. (eds.) Dynamical Systems and Their Applications in Biology. American Mathematical Society, New York (2003)
6. Bhopal, R.S.: Concepts of Epidemiology: Integrating the Ideas, Theories, Principles, and Methods of Epidemiology. Oxford University Press, Oxford (2016)
7. Box, G.E.P., Jenkins, G.M., Reinsel, G.C., Ljung, G.M.: Time Series Analysis. Forecasting and Control, 5th edn. Wiley, Hoboken (2016)
8. Bürger, R., Coronel, A., Sepúlveda, M.: Numerical solution of an inverse problem for a scalar conservation law modelling sedimentation, hyperbolic problems: theory, numerics and applications. Proc. Symp. Appl. Math. **67**, 445–454 (2009)

9. Cauchemez, S., Ferguson, N.M.: Likelihood based estimation of continuous time epidemic models from timeseries data: application to measles transmission in London. J. R. Soc. Interface **5**(25), 885–897 (2008)
10. Coronel, A., James, F., Sepúlveda, M.: Numerical identification of parameters for a model of sedimentation processes. Inverse Probl. **19**(4), 951–972 (2003)
11. Dayan, G.H., Shaw, K.M., Baughman, A.L., Orellana, L.C., Forlenza, R., Ellis, A., Chaui, J., Kaplan, S., Strebel, P.: Assessment of delay in age-appropriate vaccination using survival analysis. Am. J. Epidemiol. **163**(6), 561–570 (2006)
12. Detels, R.: Epidemiology: the foundation of public health. Oxford Textbook of Global Public Health, 5th edn. (2015). https://doi.org/10.1093/med/9780199661756.001.0001
13. Diekmann, O., Heesterbeek, J.A.: Mathematical Epidemiology of Infectious Diseases: Model Building, Analysis and Interpretation. Wiley, New York (2000)
14. Engl, H.W., Flamm, C., Kügler, P., Lu, J., Müller, S., Schuster, P.: Inverse problems in systems biology. Inverse Probl. **25**, 1–51 (2009). https://doi.org/10.1088/0266-5611/25/12/123014
15. Estrella, E.: Consecuencias epidemiológicas de la conquista de América. MS. Dpto. de Humanidades Médicas, Facultad de Medicina, Universidad Central de Quito, Ecuador (2002)
16. Frasso, G., Lambert, P.: Bayesian inference in an extended SEIR model with nonparametric disease transmission rate: an application to the Ebola epidemic in Sierra Leone. Biostatistics **17**(4), 779–792 (2016)
17. Gibson, G.J., Renshaw, E.: Estimating parameters in stochastic compartmental models using Markov chain methods. Math. Med. Biol.: J. IMA **15**(1), 19–40 (1998)
18. Hethcote, H.W.: The mathematics of infectious diseases. SIAM **42**, 599–653 (2000)
19. Last, J.M., Abramson, J.H., Freidman, G.D.: A Dictionary of Epidemiology, 6th edn. Oxford University Press, New York (2014)
20. Lekone, P.E., Finkenstädt, B.F.: Statistical inference in a stochastic epidemic SEIR model with control intervention: Ebola as a case study. Biometrics **62**(4), 1170–1177 (2006)
21. Marinova, T.T., Marinova, R.S., Omojola, J., Jackson, M.: Inverse problem for coefficient identification in SIR epidemic models. Comput. Math. Appl. **67**, 2218–2227 (2014). https://doi.org/10.1016/j.camwa.2014.02.002
22. McKinley, T.J., Ross, J.V., Deardon, R., Cook, A.R.: Simulation based Bayesian inference for epidemic models. Comput. Stat. Data Anal. **71**, 434–447 (2014)
23. Mohammadbeigi, A., Mokhtari, M., Zahraei, S.M., Eshrati, B., Rejali, M.: Survival analysis for predictive factors of delay vaccination in Iranian children. Int. J. Prev. Med. **6**(1), 119–123 (2015). https://doi.org/10.4103/2008-7802.170868
24. Norrie, P.: A History of Disease in Ancient Times: More Lethal than War. Springer, Berlin (2016)
25. Panagiotakopulu, E.: Pharaonic Egypt and the origins of plague. J. Biogeogr. **31**(2), 269–275 (2004). https://doi.org/10.1046/j.0305-0270.2003.01009.x
26. Pesco, P.: Modelos estocásticos para epidemias recurrentes. Tesis doctoral, La Plata, Argentina (2017)
27. Rhodes, A., Allman, E.S.: Mathematical Models in Biology: An Introduction. Cambridge University Press, New York (2003)
28. R Core Team: R: A language and environment for statistical computing. R Foundation for Statistical Computing, Vienna, Austria (2018). http://www.R-project.org
29. Streftaris, G., Gibson, G.J.: Statistical inference for stochastic epidemic models. In: Proceedings of the 17th International Workshop on Statistical Modeling, pp. 609–616 (2002)
30. Tarantola, A.: Inverse Problem Theory and Methods for Model Parameter Estimation. SIAM, Philadelphia (2005)
31. Wang, W., Ruan, S.: Bifurcations in an epidemic model with constant removal rate of the infectives. J. Math. Anal. Appl. **291**, 775–793 (2004). https://doi.org/10.1016/j.jmaa.2003.11.043
32. Wu, L., Feng, Z.: Homoclinic bifurcation in an SIQR model for childhood diseases. J. Differ. Equ. **168**, 150–167 (2000). https://doi.org/10.1006/jdeq.2000.3882

33. Xu, X., Kypraios, T., O'neill, P.D.: Bayesian nonparametric inference for stochastic epidemic models using Gaussian processes. Biostatistics **17**(4), 619–633 (2016)
34. Zietz, B.P., Dunkelberg, H.: The history of the plague and the research on the causative agent Yersinia pestis. Int. J. Hyg. Environ. Health **207**(2), 165–178 (2004). https://doi.org/10.1078/1438-4639-00259

Impact of the Red Code Process Using Structural Equation Models

Eduardo Pérez Castro, Flaviano Godínez Jaimes, Elia Barrera Rodríguez, Ramón Reyes Carreto, Raúl López Roque and Virginia Vera Leyva

Abstract This paper proposes an ad hoc model to explain the relationships between latent and observed variables, which influence the results of the care of pregnant woman with obstetric emergency before and after the implementation of a standardized process called *Red Code*. It has used information from medical records of pregnant women who were treated in the emergency service of the Hospital de la Madre y el Niño Guerrerense, Guerrero, Mexico. Based on expert judgment, 19 observed variables were grouped into 5 latent variables: first hemodynamic state, second hemodynamic state, obstetric-gynecological history, treatments, and results of EMOC. An ad hoc model was proposed that includes the first four latent variables as independent and the last one as a latent dependent variable. To asses the proposal, goodness-of-fit indexes for the fitted structural equation model were used. It was concluded that the results are mainly affected by obstetric-gynecological history and second hemodynamic status for the before red code period and obstetric-gynecological history and treatment for the red code period.

Keywords Red code · Structural equation models · Obstetric emergency

E. Pérez Castro (✉) · F. Godínez Jaimes (✉) · R. Reyes Carreto
Facultad de Matemáticas, Universidad Autónoma de Guerrero, Chilpancingo, Mexico
e-mail: laloperezcastro@gmail.com

F. Godínez Jaimes
e-mail: fgodinezj@uagro.mx

R. Reyes Carreto
e-mail: rrcarreto@gmail.com

E. Barrera Rodríguez (✉)
Unidad de Innovación Clínica y Epidemiológica del Estado de Guerrero,
Hospital de la Madre y el Niño Guerrerense, Chilpancingo, Mexico
e-mail: eliabare76@gmail.com

R. López Roque · V. Vera Leyva
Hospital de la Madre y el Niño Guerrerense, Chilpancingo, Mexico
e-mail: litmanlopez@hotmail.com

V. Vera Leyva
e-mail: chompyjavy@hotmail.com

I. Antoniano-Villalobos et al. (eds.), *Selected Contributions on Statistics and Data Science in Latin America*, Springer Proceedings in Mathematics & Statistics 301,
https://doi.org/10.1007/978-3-030-31551-1_9

1 Introduction

Maternal survival has been one of the priority topics of global health in recent decades. This issue was adopted as the fifth Millennium Development Goal (MDG) (Arregoces 2015; Smith 2016 cited in [1]) and later as the third Sustainable Development Goal. Despite the remarkable achievements in reducing maternal mortality, around 830 women around the world die each day from complications related to pregnancy or childbirth.

México is divided into 32 states and Guerrero is one of poorest states in the country where more women die during pregnancy, childbirth, and puerperium, mainly in the indigenous regions known as Montaña and Costa Chica. For this reason and with the aim of fulfilling the fifth MDG, in 2006, the Hospital de la Madre y el Niño Guerrerense (HMNG) begins the training of personnel in emergency obstetric care (EMOC) with the course-workshop called *Intensive and integral management of the pregnant woman in a critical state*.

According to the Official Mexican Standard 007 (NOM 007 for its acronym in Spanish), an obstetric emergency (OE) is defined as the complication or intercurrence condition of pregnancy that implies a risk of maternal–perinatal morbidity or mortality.

The main pregnancy complications that cause OE are (a) obstetric hemorrhage during pregnancy, childbirth, or after delivery, (b) hypertensive disorders in pregnancy, which are classified as preeclampsia, eclampsia, and (c) obstetric sepsis [2].

Obstetric hemorrhage is defined as the loss of blood greater than $500\,\text{cm}^3$ after a vaginal delivery or greater than $1000\,\text{cm}^3$ after a caesarean section or less than $1000\,\text{cm}^3$ if there are signs of hypovolemic shock. Preeclampsia or eclampsia occurs in 3–5% of pregnant women and it is characterized for hypertension after the 20th week of gestation. Finally, obstetric sepsis occurs when two or more of the following signs exist: temperature greater than 38 °C or less than 36 °C; heart rate greater than 90 beats/minute; breathing frequency greater than 20 beats/minute or partial pressure of carbon dioxide greater than 32 mmHg; leukocyte count greater than $12{,}000/\text{mm}^3$ or less than $4{,}000/\text{mm}^3$, with more than 10% of immature leukocyte forms; failure of distant organs where the symptomatology depends on the affected organs, and coagulation, liver, kidney, breathing, or neurological disorders may occur [3].

Red Code (RC) is the procedure aimed at pregnant women with obstetric emergencies, which include the identification of morbidity, initial management, and conditions for referral when necessary.

The RC is an organized work scheme that, when an OE occurs, provides an assistance team to work in an orderly and coordinated way in order to reduce the pregnant woman's morbidity and mortality. The RC begins when the pregnant woman in the emergency room presents a pregnancy complication that may be a maternal–perinatal risk.

To monitor the process of EMOC has proposed the assessment of indicators in order to identify the medical care components that are working well, those that need improvement, change, or further investigation [4]. In this way, the evaluation and monitoring of the EMOC indicators allow decision-makers to identify needs to plan interventions that improve the functioning of programs and services that favor the EMOC and thus contribute to the reduction of maternal and neonatal mortality.

The RC process begins with the activation of an audible alarm when health personnel detects the OE and the work team meets in the emergency room. The nurse performs the registration of general data, gynecological and obstetric history, performs somatometry, and takes vital signs such as blood pressure, temperature, heart rate, breathing frequency, and catheter placement. She also starts the administration of medications and prepares the patient for admission to the next care service.

The obstetrician performs the comprehensive maternal-fetal assessment and determines the diagnosis and management plan. At the same time, the internist performs the neurological, hemodynamic, and metabolic assessment. Together, obstetrician and internist establish medical treatment and decide what is the next healthcare service for the pregnant woman with EO. Similarly, the pediatrician assesses the maternal-fetal conditions and foresees the need to perform neonatal resuscitation. Laboratory and cabinet studies are taken by the corresponding personnel, as well as the request for hemocompatibility tests of blood components (erythrocyte package, fresh frozen plasma, platelet package) to the blood bank.

The obstetrician determines if the patient responded to the initial management. When the patient survives it is called "stabilized patient"; in case of death, it is called "failed stabilization". If the patient was stabilized, the obstetrician decides if it requires Toco-Surgical management or admission to the adult Intensive Care Unit (ICU), and the patient is immediately admitted to the second service.

As soon as the patient is admitted to the second care service, the nurse performs the vital signs registration again and continues with the medical surgical management established as the administration of drugs and blood components.

If the patient dies, the obstetrician informs the relative responsible for the patient about the consequences that conditioned the maternal death, elaborates Death Certificate and confidential maternal death questionnaire of the Secretary of Health.

On the other hand, the Social Work staff performs the record and follow-up of RC patients up to 40 days after their admission with the aim of verifying the health status of RC patients and supervising the adherence to medical treatment to avoid sequelae and maternal deaths during the quarantine. At this point, the RC ends.

Measuring the effectiveness of the services requires insight and detail, and current monitoring efforts are inconsistent. Measuring the effectiveness of the health services requires insight and detail, current monitoring efforts are inconsistent. Several researches have studied the primary healthcare facilities, the antibiotics administration, the manual removal of the placenta, and removal of retained products during the assisted vaginal delivery using a vacuum extractor. Other researchers have focused on the basic neonatal care including neonatal resuscitation, caesarean section, safe blood transfusion services, and the treatment of the sick baby. Although the Hospital del Madre y el Niño Guerrerense has been certified by ISO-9001:2018, the impact of these services on improving maternal health has not been measured yet.

There is empirical evidence that the implementation of RC has improved the care of pregnant women with obstetric emergencies. However, a formal study confirming it has not been carried out. The objective of this work is to evaluate the impact of RC implementation, using a structural equations model (SEM), for the care of the pregnant woman with OE. Because there is no model used in the literature, an ad hoc model based on expert physicians is proposed.

2 Methodology

This section presents basic theory of the SEM, the database used for the analysis is described, as well as the formulation of the ad hoc model proposed.

2.1 Structural Equation Models

The SEM permits to study the latent independent variables effect on a latent dependent variable. The latent variables cannot directly be measured but they do through observed variables that in this work are those that are involved in the EMOC. SEM has received a great deal of attention in biomedical research [5, 6]. An SEM is formed by (i) a set of measurement equations that represents the relationships of each latent variable (or factor) with their corresponding observed variables and (ii) a set of structural equations where the relationships between the latent variables are described. The measurement equations are defined by [7]

$$\boldsymbol{y} = \boldsymbol{\Lambda}\boldsymbol{\omega} + \boldsymbol{\varepsilon} \tag{1}$$

where $\boldsymbol{y} = \left(y_1, \ldots, y_p\right)$ is a $p \times 1$ random vector of observed values, $\boldsymbol{\Lambda}$ is a $p \times q$ matrix of factorial loads, $\boldsymbol{\omega}$ is a vector of latent variables, and $\boldsymbol{\varepsilon}$ is a random vector of measurement errors (residual). It is assumed that $\boldsymbol{\varepsilon}$ is distributed $N(\boldsymbol{0}, \boldsymbol{\Psi}_{\varepsilon})$ where $\boldsymbol{\Psi}_{\varepsilon}$ is a diagonal matrix. In this work, $\boldsymbol{y}$ is a 19×1 vector, $\boldsymbol{\Lambda}$ is a 19×5 matrix, $\boldsymbol{\omega}$ is a 5×1 vector, and $\boldsymbol{\varepsilon}$ is a 19×1 vector.

Let $\boldsymbol{\omega} = (\boldsymbol{\eta}^T, \boldsymbol{\xi}^T)^T$, with $\boldsymbol{\eta}$ $(q_1 \times 1)$ dependent latent vector and $\boldsymbol{\xi}$ $(q_2 \times 1)$ independent latent vector. The structural equations that define the relations between $\boldsymbol{\eta}$ and $\boldsymbol{\xi}$ are given by

$$\boldsymbol{\eta} = \boldsymbol{\Pi}\boldsymbol{\eta} + \boldsymbol{\Gamma}\boldsymbol{\xi} + \boldsymbol{\delta} \tag{2}$$

where $\boldsymbol{\Pi}$ $(q_1 \times q_1)$ and $\boldsymbol{\Gamma}$ $(q_1 \times q_2)$ are matrices of unknown regression coefficients that measure the causal effect of $\boldsymbol{\xi}$ on $\boldsymbol{\eta}$ and $\boldsymbol{\delta}$ $(q_1 \times 1)$ is a random vector of measurement errors. The dimensions of $\boldsymbol{\eta}$, $\boldsymbol{\Pi}$, $\boldsymbol{\Gamma}$, $\boldsymbol{\xi}$, and $\boldsymbol{\delta}$ are $4 \times 1, 4 \times 4, 4 \times 1, 1 \times 1$, and 4×1, respectively. The model assumes that $\boldsymbol{\xi} \sim N(\boldsymbol{0}, \boldsymbol{\Phi})$ where $\boldsymbol{\Phi}$ is a general covariance matrix and $\boldsymbol{\delta} \sim N(\boldsymbol{0}, \boldsymbol{\Psi}_{\delta})$, where $\boldsymbol{\Psi}_{\delta}$ is a diagonal matrix. In addition, $\boldsymbol{\delta}$ is independent of $\boldsymbol{\xi}$, and $\boldsymbol{\varepsilon}$ is uncorrelated with $\boldsymbol{\omega}$ and $\boldsymbol{\delta}$. It is also assumed that $(\boldsymbol{I} - \boldsymbol{\Pi})$ is nonsingular.

2.1.1 Model Estimation

The estimation involves determining the values of the unknown parameters and their respective measurement error. One of the techniques widely used in most of the statistical packages for the SEM estimation is maximum likelihood (ML) [8]. Although there are alternative methods, such as weighted least squares (WLS), generalized least squares (GLS), and asymptotically free distributed (AGL), in this work, ML was used.

2.1.2 Model Diagnostics

Some indexes were used to evaluate the fit of the model that were useful in determining how well the ad hoc model fits the sample data [1]. The goodness-of-fit index (GFI) and adjusted GFI (AGFI) represent the proportion of variance, analogous to R^2. GFI and AGFI values range between 0 and 1 and it is generally accepted that values of 0.90 or greater indicate well-fitted models. The root mean square error of approximation (RMSEA) is a badness of fit index. If the RMSEA is equal to zero, the best fit occurs, if it is less than 0.05 a good fit, and less than 0.08 an appropriate fit. The comparative fit index (CFI) is an index that compares the fit of the theoretical model with the fit of the independence model, which is the one in which all the variables are independent. The CFI values vary between 0 and 1, and a good fit is reached if values greater than 0.90 occur. This value indicates that at least 90% of the variance in the data can be reproduced by the model. Another index is the standardized root mean squared residual (SRMR) which is obtained by dividing the RMSEA by the standard deviation. This index is considered indicative of a good fit if it is less than 0.08.

2.2 Database

The RC was implemented in 2011 and since then it has been in operation. The before red code (BRC) period was defined as the time period from 2009 to 2011 and the RC period as the time period from 2013 to 2015. Although in the intermediate period the RC already operated, it was still in its adaptation phase. For this reason, this period of time was not considered.

The database corresponds to a case series study of pregnant women who attended to the HMNG's emergency service with an OE. The base consists of two groups of patients, the first group was attended from January 2009 to December 2011 and corresponds to the BRC period and another group of patients attended from September 2013 to December 2015 with the RC procedure implemented.

The list of patients was obtained from the emergency records in the studied periods, and their clinical records were used to obtain the needed information. Incomplete records were not used in the database. The final database has 230 observations for BRC and 106 for RC periods, respectively.

The first column of Table 1 lists the 19 observed variables included in the database and the 5 latent variables formed according to the judgment of expert physicians. The latent variable **First hemodynamic state** is related to the actions made by the nurse when the pregnant woman is received in the emergency service. The variables measured are temperature, heart rate, blood pressure, breathing frequency, and number of convulsions. In the same place, a nurse gets information about the second latent variable **Obstetric-gynecological history** that is measured by the observed variables such as number of abortions, number of caesarean sections, pregnant's weight, and number of vaginal deliveries. Physicians determine the **Treatments** needed to face the OE that can include platelets, plasma, and erythrocyte concentrate. The **Second hemodynamic state** corresponds to the actions taken when the pregnant woman is transferred from the emergency service to the ICU or Toco-surgery. The variables measured are the same observed variables for the first hemodynamic state except for the number of convulsions. The latent variable **Results of the emergency obstetric care** measures the consequences of the actions made in the RC process. The observed variables are the number of sequelae, the newborn's weight, and the gestation weeks.

2.3 Ad hoc Model

The proposed ad hoc model, Eq. (3), was proposed based on the knowledge of expert physicians. The proposed ad hoc SEM consists of five latent variables: four latent dependent variables η_1, η_2, η_3, and η_4 and one latent independent variable, ξ. Table 1 shows the factors considered and the observed variables that comprise it as well as its mathematical notation. The model assumes that

1. The latent variable first hemodynamic state is measured through the observed variables: temperature, heart rate, blood pressure, breathing frequency, and number of convulsions.
2. The latent variable second hemodynamic state is measured by the observed variables: heart rate, blood pressure, breathing frequency, and temperature.
3. The latent variable obstetric-gynecological history is measured by the observed variables: number of abortions, number of caesarean sections, pregnant's weight, and number of vaginal deliveries.
4. The latent variable Treatments is measured by the observed variables: platelets, plasma, and erythrocyte concentrate.
5. The latent variable Results of the emergency obstetric care is measured by the observable variables: number of sequelae, newborn's weight, and gestation weeks.
6. The obstetric-gynecological history has an effect on the first hemodynamic state.
7. The first hemodynamic state has an effect on the second hemodynamic state.
8. The first and second hemodynamic states and the obstetric-gynecological history have effect on the treatments.
9. The first and second hemodynamic states, the obstetric-gynecological history, and the treatments have effect on the results of the emergency obstetric care.

The measurement Eqs. (1) become Eqs. (3) defined by using 19 observable variables $\boldsymbol{y_i} = (y_{i1}, \ldots, y_{i,19})$ and 5 latent variables $\boldsymbol{\omega} = (\eta_1, \eta_2, \eta_3, \eta_4, \xi)^T$ as follows:

$$\begin{bmatrix} Tm1 \\ Hr1 \\ Bp1 \\ Bf1 \\ Nc \\ Hr2 \\ Bp2 \\ Bf2 \\ Tm2 \\ Na \\ Ncs \\ Pw \\ Nvd \\ Pla \\ Pls \\ Ec \\ Nms \\ Nw \\ Gw \end{bmatrix} = \begin{bmatrix} \lambda_{11} & 0 & 0 & 0 & 0 \\ \lambda_{21} & 0 & 0 & 0 & 0 \\ \lambda_{31} & 0 & 0 & 0 & 0 \\ \lambda_{41} & 0 & 0 & 0 & 0 \\ \lambda_{51} & 0 & 0 & 0 & 0 \\ 0 & \lambda_{62} & 0 & 0 & 0 \\ 0 & \lambda_{72} & 0 & 0 & 0 \\ 0 & \lambda_{82} & 0 & 0 & 0 \\ 0 & \lambda_{94} & 0 & 0 & 0 \\ 0 & 0 & \lambda_{10,3} & 0 & 0 \\ 0 & 0 & \lambda_{11,3} & 0 & 0 \\ 0 & 0 & \lambda_{12,3} & 0 & 0 \\ 0 & 0 & \lambda_{13,3} & 0 & 0 \\ 0 & 0 & 0 & \lambda_{14,4} & 0 \\ 0 & 0 & 0 & \lambda_{15,4} & 0 \\ 0 & 0 & 0 & \lambda_{16,4} & 0 \\ 0 & 0 & 0 & 0 & \lambda_{17,5} \\ 0 & 0 & 0 & 0 & \lambda_{18,5} \\ 0 & 0 & 0 & 0 & \lambda_{19,5} \end{bmatrix} \begin{bmatrix} FHS \\ SHS \\ OGH \\ Treat \\ REMOC \end{bmatrix} + \begin{bmatrix} \varepsilon_1 \\ \varepsilon_2 \\ \varepsilon_3 \\ \varepsilon_4 \\ \varepsilon_5 \\ \varepsilon_6 \\ \varepsilon_7 \\ \varepsilon_8 \\ \varepsilon_9 \\ \varepsilon_{10} \\ \varepsilon_{11} \\ \varepsilon_{12} \\ \varepsilon_{13} \\ \varepsilon_{14} \\ \varepsilon_{15} \\ \varepsilon_{16} \\ \varepsilon_{17} \\ \varepsilon_{18} \\ \varepsilon_{19} \end{bmatrix} \quad (3)$$

Similar to confirmatory factor analysis, one the factor loadings must be fixed in each latent variable in order to fit the model. The fixed loads were $\lambda_{11} = \lambda_{62} = \lambda_{10,3} = \lambda_{14,4} = \lambda_{17,5} = 1$.

The structural Eqs. (2) become Eqs. (4), and its matrix formulation is

$$\begin{bmatrix} FHS \\ SHS \\ Treat \\ REMOC \end{bmatrix} = \begin{bmatrix} 0 & 0 & 0 & 0 \\ \pi_{21} & 0 & 0 & 0 \\ \pi_{31} & \pi_{32} & 0 & 0 \\ \pi_{41} & \pi_{42} & \pi_{43} & 0 \end{bmatrix} \begin{bmatrix} FHS \\ SHS \\ Treat \\ REMOC \end{bmatrix} + \begin{bmatrix} \gamma_{11} \\ 0 \\ \gamma_{31} \\ \gamma_{41} \end{bmatrix} OGH + \begin{bmatrix} \delta_1 \\ \delta_2 \\ \delta_3 \\ \delta_4 \end{bmatrix} \quad (4)$$

Two SEM models were fitted, the first one to the BRC period and the second one to the RC period. The *sem* function on the library *laavan* [9] implemented in the R package [10] was used to fit the models. The analysis was done using the correlation matrix because the variables have very different variances.

3 Results

In this section, the estimation of the SEM, the goodness-of-fit indexes of the models in both periods BRC and RC, and the interpretations of the fitted models are shown.

Table 1 Exploratory analysis of the variables studied for the BRC and RC periods

Variables		BRC			RC			p-value
		Mean	Sd*	Median	Mean	Sd*	Median	
First hemodynamic state (FHS) η_1								
Temperature (Tm1)	y_1	36.41	0.45	36.40	36.24	0.49	36.40	0.00
Heart rate (Hr1)	y_2	83.69	12.04	81.00	84.92	123.00	81.00	0.33
Blood pressure(Bp1)	y_3	139.79	21.71	140.00	152.77	24.31	140.00	0.00**
Breathing frequency (Bf1)	y_4	20.96	2.81	20.00	21.96	2.41	20.00	0.00
Number of convulsions (Nc)	y_5	0.10	0.83	0.00	0.06	0.36	0.00	0.50
Second hemodynamic state (SHS) η_2								
Heart rate (Hr2)	y_6	87.53	14.76	84.00	95.11	15.54	84.00	0.00**
Blood pressure (Bp2)	y_7	85.98	17.94	84.50	85.87	13.66	84.50	0.81
Breathing frequency(Bf2)	y_8	21.73	5.08	21.00	21.85	2.58	21.00	0.78
Temperature (Tm2)	y_9	36.44	0.46	36.50	36.22	1.08	36.50	0.04
Obstetric-gynecological history (OGH) ξ								
Number of abortions (Na)	y_{10}	0.20	0.49	0.00	0.18	0.77	0.00	0.83
Number of caesareans section (Ncs)	y_{11}	0.21	0.50	0.00	0.14	0.42	0.00	0.20
Pregnant's weight (Pw)	y_{12}	69.78	13.82	68.00	68.79	12.99	68.00	0.52
Number of vaginal deliveries (Nvd)	y_{13}	1.41	2.30	0.00	1.24	2.04	0.00	0.47
Treatments (Treat) η_3								
Platelets (Pla)	y_{14}	0.26	1.49	0.00	0.01	0.10	0.00	0.01
Plasma (Pls)	y_{15}	0.27	1.48	0.00	0.04	0.04	0.00	0.01
Erythrocyte concentrate (Ec)	y_{16}	0.44	1.25	0.00	0.23	0.76	0.00	0.05
Results of the emergency obstetric care (REMOC) η_4								
Number of sequelae (Nms)	y_{17}	0.17	0.45	0.00	0.05	0.21	0.00	0.00
Newborn's weight (Nw)	y_{18}	2554.47	712.97	2757.50	2706.81	708.99	2757.50	0.06
Gestation weeks (Gw)	y_{19}	37.48	3.04	38.25	37.81	2.60	38.25	0.28

*Standard deviation, **p-value $< 1 \times 10^{-6}$

3.1 Exploratory Analysis

Table 1 shows the means, the medians, and standard deviations of the variables studied. It also displays the p-value of the null hypothesis of equality of the population means in BRC and RC periods using the t test for independent samples assuming variances different or equal populations as appropriate.

Some important results that are observed in Table 1 are mentioned. According to the literature, convulsions in a pregnant woman in addition to hypertension are clear signs of eclampsia, which is one of the main causes of OE. It is observed that a significative decrease in the number of convulsions during the RC period indicates that the RC has a positive impact improving the maternal health. Another significative decrease occurred in the use of blood products such as erythrocyte concentrate, platelet, and plasma in the treatment of pregnant women with OE during the RC period. Finally, the decrease in the number of sequelae, as well as the increase in the newborn's weight in the BRC period, shows the impact on improving the health of the child–mother binomial.

Correlations among the observed variables are shown in Fig. 1. Blue shading indicates positive correlations and red shading indicates negative correlations. The p-value indicates that the population correlation is equal to zero. This was calculated using a t test. In the BRC period, there are variables with high negative correlations, for example, erythrocyte concentrate-pregnant's weight ($r = -0.17$, $p = 0.02$) and gestation weeks-number of caesareans section ($r = -0.17, p = 0.01$). There are also variables with high positive correlations like gestation weeks-newborn's weight ($r = 0.67$, $p < 0.00$), platelets-number of sequelae ($r = 0.31$, $p < 0.00$), and plasma-platelets ($r = 0.87$, $p < 0.00$). In the RC period, there were low correlations among the variables, some pairs of variables with high correlations are platelets-number of sequelae ($r = 0.43$, $p < 0.00$) and number of vaginal deliveries-plasma ($r = 0.30$, $p < 0.00$).

3.2 Model Estimation

Results of the fitted model to both, BRC and RC, periods are shown in Tables 3 and 4. However, before interpreting the fitted model, it must be verified that the fit is appropriate.

3.3 Model Diagnostics

Goodness-of-fit indexes of the ad hoc model for the two periods are shown in Table 2. The chi-square statistic value and its p-value indicate that data are not well fitted by the ad hoc model in the BRC period; however, it is known that this statistic is affected by large sample sizes (greater than 150). In contrast, this statistic indicates a good fit of the model to the data in the RC period.

In the BRC period, the fitted model has $GFI = 0.89$ and in the RC period, it has $GFI = 0.87$ and these values are very close to the reference value 0.90, which indicates a good fit of the model. That is, 89 of variance in the observed data is in the BRC period and 87 in the RC period is explained by the ad hoc model.

In the BRC period, the fitted model has $AGFI = 0.85$ and $AGFI = 0.82$. These values are close to the reference value of 0.90, which means a good fit of the model.

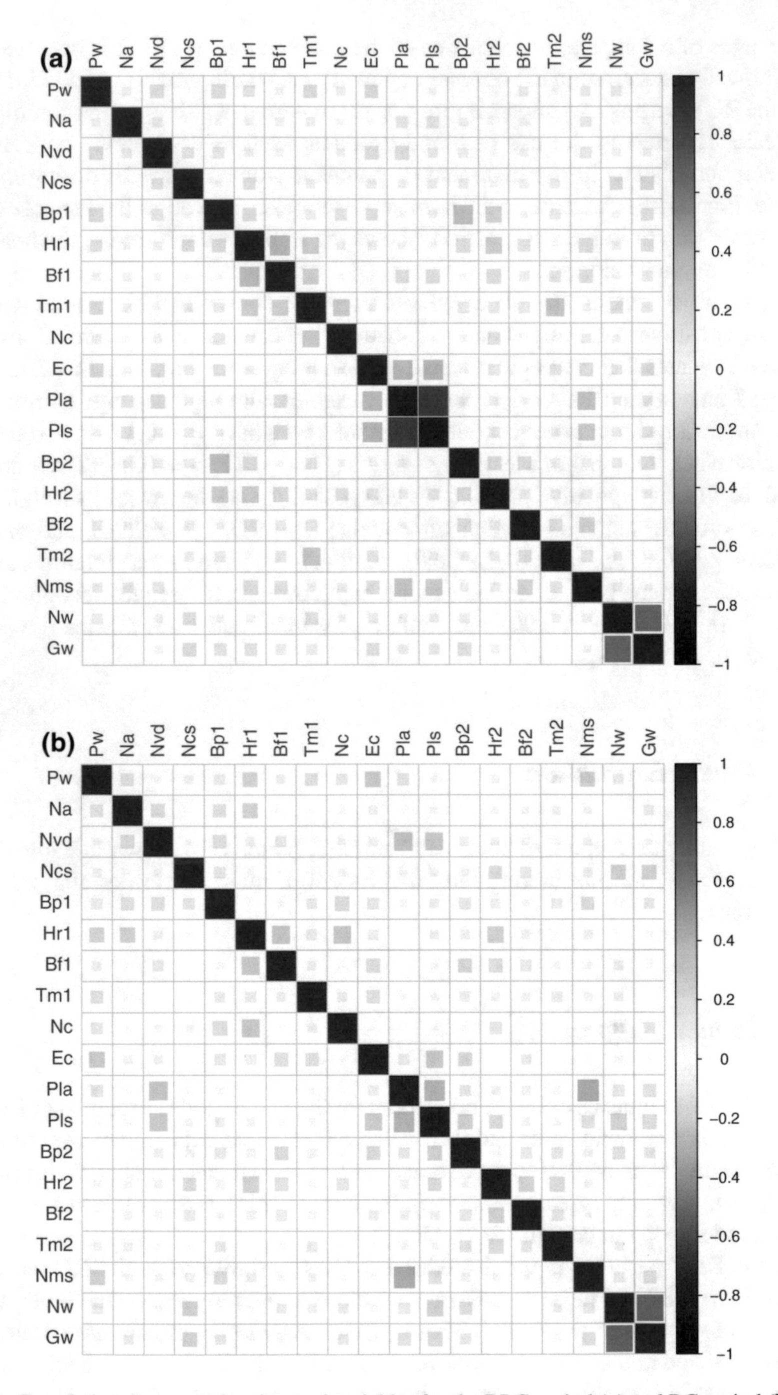

Fig. 1 Correlations between the observed variables for the BRC period (**a**) and RC period (**b**)

Table 2 Goodness-of-fit indexes of the ad hoc model for the two periods

	BRC	RC	Reference
Chi	295.5	159.8	
	$p = 0.00$	$p = 0.22$	>0.05
GFI	0.890	0.870	>0.90
AGFI	0.852	0.820	>0.90
RMSEA	0.068	0.029	<0.07
SRMR	0.078	0.080	<0.08
CFI	0.803	0.915	>0.90

They say that 85 and 82% of the population variance are explained by the ad hoc model in the BRC and RC periods, respectively.

The RMSEA value in the BRC period is 0.068 that is near the reference value 0.070 [11]. This RMSEA value indicates that the measurement model and the covariances structure of the observed variables have an appropriate fit. However, in the RC period, the RMSEA value is 0.029 that is less than 0.070, and therefore in the RC period the model has better fit between the data and the model.

The SRMR in the BRC period was 0.078 which is a little lower than the reference value 0.08 [12], and this means good fit of the model. For the RC period, the value was 0.08, which is equal to the reference value suggested in the literature.

Finally, in the BRC period, the CFI value was 0.80, close but less than the reference value 0.90 [13]. This means that the measurement model and covariance structure of the observed variables have a reasonable fit, but do not achieve the minimum acceptable. Better results occur in the RC period where the CFI is 0.915 that is greater than the reference value 0.90 which indicates a good fit between the data and the model.

In summary, the goodness-of-fit indexes for the fitted SEM in the two periods, BRC and RC, vary from reasonable to good, and thus the fitted models can be interpreted.

3.4 Model Inference

Parameter estimation of the ad hoc SEM proposed for the two periods is presented in Tables 3 and 4. The observed variables that have more impact on each latent variable in the measurement model are as follows:

- First hemodynamic state: heart rate in both periods BRC and RC (1.73 and 4.38, respectively).
- Second hemodynamic state: heart rate in both periods BRC and RC (1.00 in both periods).

Table 3 Measurement model estimates

	BRC	RC		BRC	RC
First hemodynamic state			Obstetric-gynecological history		
Blood pressure	1.00*	1.00	Pregnant's weight	1.00	1.00
Breathing frequency	1.41*	−1.75	Number of vaginal deliveries	2.57	0.36
Heart rate	1.73*	4.38	Number of abortions	1.73	−6.77
Temperature	1.33*	0.10	Number of caesareans section	−0.98	30.80
Number of convulsions	0.55*	−1.97			
Second hemodynamic state			Treatments		
Heart rate	1.00*	1.00	Platelets	1.00*	1.00
Temperature	0.49*	0.49	Plasma	0.95*	1.48
Blood pressure	0.80*	0.20	Erythrocyte concentrate	0.40*	0.53
Breathing frequency	0.56*	0.65			
Results of the emergency obstetric care					
Number of sequels	1.00*	1.00			
Newborn's weight	−13.84*	5.65			
Gestation weeks	−18.24*	5.13			

* p-value < 0.05

Table 4 Structural fitted model

Latent variables	BRC			RC		
	Estimate	Sd*	p-value	Estimate	Sd*	p-value
First hemodynamic state						
Obstetric-gynecological history	0.11	0.21	0.60	−1.27	2.10	0.54
Second hemodynamic state						
First hemodynamic state	1.22	0.37	0.00	−0.25	0.28	0.37
Treatments						
First hemodynamic state	−0.18	1.50	0.90	−0.17	0.39	0.65
Second hemodynamic state	0.35	1.45	0.75	−0.25	0.13	0.06
Obstetric-gynecological history	1.19	0.72	0.10	−4.24	2.90	0.53
Results of the emergency obstetric care						
Obstetric-gynecological history	−0.04	0.06	0.52	0.47	0.85	0.57
First hemodynamic state	−0.06	0.12	0.61	−0.05	0.07	0.43
Second hemodynamic state	0.01	0.07	0.80	0.01	0.03	0.72
Treatments	0.01	0.01	0.48	0.08	0.05	0.08

*Standard deviation

- Obstetric-gynecological history: number of vaginal deliveries in the BRC period and number of caesarean sections in the RC period (2.57 and −6.77, respectively).
- Treatments: platelets in the BRC period and plasma in the RC period (1.00 and 1.48, respectively).
- Results of the emergency obstetric care: gestation weeks during the BRC period and newborn's weight during the RC period (−18.24 and 5.65, respectively).

The interpretation of these estimates is as follows, in the BRC period, when the first hemodynamic state increases one unit, then blood pressure (Bp1), breathing frequency (Bf1), heart rate (Hr1), number of convulsions (Nc), and temperature (Tm1) increase 1.00, 1.41, 1.73, 0.55, and 1.33 units, respectively. On the other hand, in the RC period, when first hemodynamic state increases one unit, then blood pressure (Bp1), heart rate (Hr1), temperature (Tm1) increase 1.00, 4.38, and 0.10, respectively, and breathing frequency (Bf1) and number of convulsions (Nc) decrease 1.75 and 1.97 units. Similar interpretations can be made for the other factors latent variables.

The factor loads corresponding to the breathing frequency had a change of sign. In the BRC period, it had a positive sign and in the RC period, it was negative. This shows the protector effect of the RC, and the same happens with the number of convulsions. In addition, the factorial loads of the newborn's weight and gestation weeks have a negative sign in the BRC and positive in the RC period. That is, if the results of the EMOC increases by a unit, in the BRC periods, newborn's weight decreases but in the RC period the weight increases. In other words, babies were to be born with low weight and were premature during the BRC period and the opposite during the RC.

The main causes of the OE in both periods are the hypertensive disorders of pregnancy. The results show that early and opportune management of the pregnant woman with OE reduces the risk of eclampsia, reflected in the decrease in the number of convulsions, and therefore of maternal death.

Although estimations for the structural model were not statistical significative, results in Table 4 permit to say that:

1. The effect of the obstetric-gynecological history in the first hemodynamic state is different in both periods. If the obstetric-gynecological history increases one unity then the first hemodynamic state increases 0.1 units in the BRC, but decreases 1.27 units in the RC period, that is, pregnant women treated during the RC period improve their first hemodynamic status.
2. The effect of the first hemodynamic state in the second is different in both periods. When the first hemodynamic state increases one unity then the second hemodynamic state increases 1.22 units in the BRC, but decreases 0.25 units in the RC period, that is, pregnant women treated during the RC period improve their second hemodynamic status.
3. The effect of the first and second hemodynamic states and the obstetric-gynecological history in the treatments used is different in both periods for the second and third latent variables. When the second hemodynamic state increases one unity then the treatments used increases 0.18 units in the BRC, but decreases

0.17 units in the RC period, that is, pregnant women treated during the RC period use less treatments. On the other hand, if the obstetric-gynecological history increases one unity then the treatments used increase 1.19 units in the BRC, but decreases 4.24 units in the RC period, that is, pregnant women in the RC period use significatively less treatments.

4. The effect of the first and second hemodynamic states, the obstetric-gynecological history, and the treatments used in the results of the EMOC are different in both periods only for the obstetric-gynecological history. In the RC period, if the obstetric-gynecological history increases by a unit then the results of the EMOC have an increment of 0.47 units, that is, the RC permits good control of the adverse complications immersed in the obstetric-gynecological history. Similar positive effects are observed for the first hemodynamic state and treatments used in the RC period.
5. In the BRC period, the latent variables that most impact the results of the EMOC are obstetric-gynecological history and first hemodynamic state (−0.06 and −0.04) and in the RC period, they are obstetric-gynecological history and treatments used (0.47 and 0.08).

4 Conclusions

To the best knowledge, there are no papers regarding the impact of the *red code* process. The structural equations model to measure the impact of the implementation of the standardized process *red code* was adequate and allowed to identify its positive effect in the care of the pregnant woman with obstetric emergency attended in the Hospital de la Madre y el Niño Guerrerense de Chilpancingo, Guerrero México.

A contribution of this study is a proposal to evaluate the effect of the attention of the obstetric emergency. But the most important contribution, it is to give evidence that the *red code* process helps to guarantee quality care and safety of pregnant women with obstetric emergency and thereby reduce maternal and perinatal morbidity and mortality.

Important differences emerge from the parameter estimations of the fitted model to the two periods. It can be seen a positive effect of the *red code* process in the care of pregnant women with obstetric emergency and this is of clinical importance. It has been reported that hypertensive disorders such as preeclampsia–eclampsia, postpartum hemorrhage as well as puerperal infections are the three acute maternal complications that increase the risk of having adverse results. The obstetric emergency care protocol through the process *red code* is a tool that improves attention and decreases adverse results in the child–mother binomial.

In summary, there was a positive effect on the health status of the patients treated with the RC process compared to the patient who was not.

The results of this study provide information that allows feedback and reinforce hospital management strategies in pregnant women with extreme morbidity who are treated in the Hospital de la Madre y el Niño Guerrerense and, therefore, improve the quality of the service provided.

References

1. Owili, P.O., Muga, M.A., Mendez, B.R., Chen, B.: Quality of maternity care and its determinants along the continuum in Kenya: a structural equation modeling analysis. PloS One **12**(5), e0177756 (2017)
2. Chebbo, A., Tan, S., Kassis, C., Tamura, L., Carlson, R.W.: Maternal sepsis and septic shock. Crit. Care Clin. **32**(1), 119–135 (2016)
3. American College of Obstetricians, Gynecologists, et al.: Hypertension in pregnancy. Report of the American college of obstetricians and gynecologists task force on hypertension in pregnancy. Obstet. Gynecol. **122**(5), 1122–1131 (2013)
4. World Health Organization, UNICEF, et al.: Monitoring emergency obstetric care: a handbook (2009)
5. Romero-Ibarguengoitia, M.E., Vadillo-Ortega, F., Caballero, A.E., Ibarra-González, I., Herrera-Rosas, A., Serratos-Canales, M.F., León-Hernández, M., González-Chávez, A., Mummidi, S., Duggirala, R., et al.: Family history and obesity in youth, their effect on acylcarnitine/aminoacids metabolomics and non-alcoholic fatty liver disease (NAFLD). Structural equation modeling approach. PloS One **13**(2), e0193138 (2018)
6. Ae Ri, J., Jang, K.S.: Structural equation modeling on health-related quality of life of patients with ankylosing spondylitis. Iran. J. Public Health **46**(10), 1338–1346 (2017)
7. Lee, S.-Y., Song, X.-Y.: Basic and Advanced Bayesian Structural Equation Modeling: With Applications in the Medical and Behavioral Sciences. Wiley, New York (2012)
8. Kline, R.B.: Principles and Practice of Structural Equation Modeling. Guilford Publications, New York (2015)
9. Rosseel, Y.: Lavaan: an R package for structural equation modeling and more. Version 0.5–12 (beta). J. Stat. Softw. **48**(2), 1–36 (2012)
10. R Core Team: R: a language and environment for statistical computing. R Foundation for Statistical Computing, Vienna, Austria (2017)
11. Steiger, J.H.: Understanding the limitations of global fit assessment in structural equation modeling. Personal. Individ. Differ. **42**(5), 893–898 (2007)
12. Hu, L., Bentler, P.M.: Cutoff criteria for fit indexes in covariance structure analysis: conventional criteria versus new alternatives. Struct. Equ. Model.: Multidiscip. J. **6**(1), 1–55 (1999)
13. Bentler, P.M.: Comparative fit indexes in structural models. Psychol. Bull. **107**(2), 238–246 (1990)

On a Construction of Stationary Processes via Bilateral Matrix-Exponential Distributions

Luz Judith R. Esparza

Abstract In this paper, we consider a construction of Markov processes with invariant Bilateral Matrix-Exponential distributions. These distributions have support on the entire real line and have rational moment-generating functions, features of importance in the area of stochastic models. The approach taken is based on a latent representation of the corresponding transition probabilities. The structure of the construction goes from the particular to the general: first, we consider Erlang and Gamma distributions, and later we consider Matrix-Exponential distributions. We include a simulation study.

Keywords Bilateral matrix · Markov process

1 Introduction

In 2002, Pitt et al. [15] introduced an approach to construct strictly stationary time series models with arbitrary but given marginal distributions. Later, in 2009, Mena and Walker [12] based on that, constructed continuous time stationary Markov models using a latent representation of the corresponding transition probabilities.

In this paper, we will use this idea considering Bilateral Matrix-Exponential (BME) distributions as marginal distributions. This class of distributions was defined in [9] as a generalization of the Matrix-Exponential (ME) class (see, e.g., [2, 6, 7]). A random variable is BME distributed if its moment-generating function is rational, while a random variable is ME distributed if its Laplace transform is rational. In the last decade, these classes of distributions have become very important in the field of stochastic models. Their applicability in areas like genetics, computer science, queuing theory, finance, social science, and health, among others, has increased its importance in the study of stochastic processes.

L. J. R. Esparza (✉)
Cátedra CONACyT, Universidad Autónoma Chapingo, Texcoco, Mexico
e-mail: judithr19@gmail.com

I. Antoniano-Villalobos et al. (eds.), *Selected Contributions on Statistics and Data Science in Latin America*, Springer Proceedings in Mathematics & Statistics 301,
https://doi.org/10.1007/978-3-030-31551-1_10

Now, suppose that we want to build up a Markovian model $\{X_n\}_{n\geq 1}$ with the requirement that its marginal distribution belongs to a given parametric family, say on the form $\pi_X(x)$. The approach consists in defining this process by constructing the transition probabilities that govern it in such a way that the desired marginal remains invariant through the time.

Once the marginal form has been chosen, the construction of the transition probabilities is performed by imposing certain dependence through a latent variable with conditional density given by $f_{T|X}(t|x)$. This conditional density is used to construct the transition distribution, driving the process $\{X_n\}_{n\geq 1}$, with transition density given in the following form:

$$p(x_{n-1}, x_n) = \int f_{X|T}(x_n|t) f_{T|X}(t|x_{n-1})\eta(dt),$$

where $f_{X|T}(x|t) \propto f_{T|X}(t|x)\pi_X(x)$, and η denotes certain reference measure, in practice the Lebesgue or counting measure.

Therefore, following this methodology, we will construct stationary processes having Bilateral Matrix-Exponential distributions as marginals.

The remainder of the paper is organized as follows. In Sect. 2, we give a background of the Matrix-Exponential distributions in order to introduce the Bilateral Matrix-Exponential distributions. In Sect. 3, we will construct stationary processes using Gamma distributions. We consider both the univariate case and the multivariate case. In Sect. 4, as a generalization, we will construct stationary processes using ME distributions. Finally, in Sect. 5, we conclude the paper.

2 Matrix-Exponential Distributions

Let us consider the distributions of nonnegative random vectors with a joint rational Laplace transform, i.e., a fraction between two multidimensional polynomials. In the univariate case, these distributions are known as Matrix-Exponential (ME) distributions, since their densities can be written as linear combinations of the elements in the exponential of a matrix.

Matrix-Exponential distributions [2, 8] are a generalization of the Phase-type (PH) distributions [13, 14], for which a probabilistic interpretation is a priori more clear. ME distributions deserve attention from researchers for a number of reasons. First, ME distributions are useful in the analysis of stochastic models and, as it was proved in [2, 5], they can be used in the analysis of renewal processes and queueing systems. Second, the class of ME distributions includes all PH distributions and all Coxian distributions, highly applied in areas like finance, communication, and survival analysis, among others.

Consider a nonnegative random variable Y with rational Laplace transform, i.e.,

$$L_Y(s) = \mathbb{E}(e^{-sY}) = \frac{p(s)}{q(s)}, \quad s \geq 0,$$

where $p(s)$ and $q(s)$ are polynomials.

Definition 2.1 A nonnegative random variable Y is said to have a ME distribution if its Laplace transform $L(s) = \mathbb{E}(\exp(-sY))$ is a rational function in s.

The Laplace transform of Y can be determined from a representation $(\boldsymbol{\gamma}, \boldsymbol{L}, \boldsymbol{\ell})$ as

$$L(s) = \boldsymbol{\gamma}(s\mathbf{I} - \boldsymbol{L})^{-1}\boldsymbol{\ell}, \tag{1}$$

where, for finite $m \geq 1$, $\boldsymbol{\gamma}$ is a $1 \times m$ row vector, $\boldsymbol{L}$ is a $m \times m$ matrix, and $\boldsymbol{\ell}$ is a $m \times 1$ column vector, all with possibly complex entries, and $\mathbf{I}$ denotes the identity matrix of appropriate dimension.

As a generalization of these distributions, we will consider a wider class of distributions whose support is the whole real line.

Definition 2.2 ([9]) A random variable X is Bilateral Matrix-Exponentially (BME) distributed, if it has a rational moment-generating function $M_X(s) = \mathbb{E}(e^{sX}) = \frac{p^*(s)}{q*(s)}$, where $p^*(s)$ and $q^*(s)$ are polynomials.

As the ME representation, a BME random variable has a particular representation. We write $X \sim BME(\boldsymbol{\alpha}_+, \boldsymbol{T}_+, \boldsymbol{t}_+, \boldsymbol{\alpha}_-, \boldsymbol{T}_-, \boldsymbol{t}_-)$ when X has the density given by $f_X(x) = \boldsymbol{\alpha}_+ e^{\boldsymbol{T}_+ x}\boldsymbol{t}_+ \mathbf{1}_{\{x>0\}} + \boldsymbol{\alpha}_- e^{\boldsymbol{T}_- x}\boldsymbol{t}_- \mathbf{1}_{\{x<0\}}$, where $\boldsymbol{\alpha}_+$ is a row vector of some dimension m_+, $\boldsymbol{T}_+$ is a matrix of dimension $m_+ \times m_+$, and $\boldsymbol{t}_+$ is an m_+-dimensional column vector. Similarly, both the vectors $\boldsymbol{\alpha}_-$, $\boldsymbol{t}_-$ and the matrix $\boldsymbol{T}_-$ are defined by some dimension m_-. Without loss of generality, we can take $\boldsymbol{\alpha}_+, \boldsymbol{\alpha}_-, \boldsymbol{T}_+$, and $\boldsymbol{T}_-$ real valued such that $0 \leq \boldsymbol{\alpha}_+\boldsymbol{e} + \boldsymbol{\alpha}_-\boldsymbol{e} \leq 1$, and $\boldsymbol{T}_+\boldsymbol{e} + \boldsymbol{t}_+ = \boldsymbol{T}_-\boldsymbol{e} + \boldsymbol{t}_- = 0$, where $\boldsymbol{e}$ is a vector of ones.

We know that if $X \sim BME(\boldsymbol{\alpha}_+, \boldsymbol{T}_+, \boldsymbol{t}_+, \boldsymbol{\alpha}_-, \boldsymbol{T}_-, \boldsymbol{t}_-)$, then its moment-generating function is given by

$$M_X(s) = (1 - \boldsymbol{\alpha}_+\boldsymbol{e} - \boldsymbol{\alpha}_-\boldsymbol{e}) + \boldsymbol{\alpha}_+(-s\boldsymbol{I} - \boldsymbol{T}_+)^{-1}\boldsymbol{t}_+ + \boldsymbol{\alpha}_-(-s\boldsymbol{I} - \boldsymbol{T}_-)^{-1}\boldsymbol{t}_-. \tag{2}$$

We recommend the reader to check [9] for more details.

On the other hand, we can interpret a PH random variable as resulting from a simple reward structure on a finite Markov jump process. If the reward rate is 1 in each state, then the total reward is PH distributed (see [1, 10]). Therefore, considering the PH class the rewards are strictly positive. As a generalization, BME distributions consider positive and negative rewards, this feature makes the BME class has applications in areas like statistics, finance, and computer science, where general reward rates may have advantages. For example, in [9], the authors applied this class of distributions considering Markov additive processes with absorption.

In the following section, we will construct processes with BME as marginal distributions.

3 Construction Using Erlang Distributions

In this section, we will construct processes through the Erlang distributions. These distributions belong to the Phase-type class. We will consider both the univariate and multivariate cases.

3.1 Univariate Case

For the univariate case, we will consider the following three cases:

1. Let $f_{X|T} = N(0, \sigma^2 T)$, where $\sigma > 0$ and T is exponentially distributed, i.e., $T \sim \exp(\lambda)$, $\lambda > 0$.
2. Considering $f_{X|T} = N(0, T)$ and T is Gamma distributed, i.e., $T \sim Ga(\lambda, \alpha^2/2)$ for $\lambda, \alpha > 0$.
3. And finally, the third case is considering $f_{X|T} = N(\beta T, T)$ and $T \sim Ga(\lambda, (\alpha^2 - \beta^2)/2)$, where $\lambda > 0$ and $\alpha^2 - \beta^2 > 0$.

3.1.1 First Case

Let T be a random variable with exponential distribution with intensity $\lambda > 0$ and moment-generating function (mgf) given by

$$M_T(s) = \mathbb{E}(e^{sT}) = \frac{\lambda}{\lambda - s}.$$

Now, define $f_{X|T} = N(0, \sigma^2 T)$. The mgf of X is given by

$$\begin{aligned} M_X(s) &= \int_0^\infty M_{X|T}(s) dF_T = \int_0^\infty e^{\frac{1}{2}\sigma^2 T s^2} dF_T = M_T\left(\frac{1}{2}\sigma^2 s^2\right) \\ &= \frac{\lambda}{\lambda - \frac{1}{2}\sigma^2 s^2}. \end{aligned} \tag{3}$$

Note that (3) is a rational function in s, i.e., X is bilateral matrix-exponentially distributed, denoted by $X \sim BME\left(\frac{1}{2}, -\frac{\sqrt{2\lambda}}{\sigma}, \frac{\sqrt{2\lambda}}{\sigma}, \frac{1}{2}, \frac{\sqrt{2\lambda}}{\sigma}, -\frac{\sqrt{2\lambda}}{\sigma}\right)$.

Considering the case when $\sigma^2 = 1$ and $\lambda = 1$, i.e., $f_T = Exp(1)$ and $f_{X|T} = N(0, T)$, the mgf of X is given by

$$M_X(s) = \frac{1}{1 - \frac{1}{2}s^2}. \tag{4}$$

On the other hand, Bayes' theorem implies

$$\begin{aligned} f_{T|X} \propto f_{X|T} \cdot f_T &\propto T^{-1/2} \exp\left(-\frac{1}{2T}x^2\right) \cdot \exp(-T) \\ &= T^{-1/2} \exp\left(-\frac{1}{2}(x^2 T^{-1} + 2T)\right) \\ &= GIG(T; \lambda^* = 1/2, \delta^* = x^2, \gamma^* = 2), \end{aligned}$$

where GIG denotes the Generalized Inverse Gaussian parametrized as in [11].

The one-step transition probability function is given through

$$\begin{aligned} p(x_t, x_{t+1}) &= \int_0^\infty f_{X_{t+1}|T} \cdot f_{T|X_t} dT \\ &= \int_0^\infty N(X_{t+1}; 0, T) \cdot GIG(T; \lambda^* = 1/2, \delta^* = x_t^2, \gamma^* = 2) dT \\ &= GH(x_{t+1}; 1/2, \sqrt{2}, 0, x_t, 0), \end{aligned}$$

where GH denotes the Generalized Hyperbolic distribution parametrized as in [4].

Notice that, for this transition density, we have

$$f_X(x_{t+1}) = \int_{\mathbb{R}} p(x_t, x_{t+1}) dF_X(x_t),$$

where the marginal distribution is given by $f_X = GH(\lambda = 1, \alpha = \sqrt{2}, \beta = 0, \delta = 0, \mu = 0)$.

It is well known the mgf of $X \sim GH(\lambda, \alpha, 0, 0, 0)$ is given by

$$\begin{aligned} M_X(s) &= \left(\frac{\alpha^2}{\alpha^2 - s^2}\right)^{\lambda/2} \frac{K_\lambda(\delta\sqrt{\alpha^2 - s^2})}{K_\lambda(\delta\sqrt{\alpha^2})} = \left(\frac{\alpha^2}{\alpha^2 - s^2}\right)^{\lambda/2} \left(\frac{\alpha^2}{\alpha^2 - s^2}\right)^{\lambda/2} \\ &= \left(\frac{\alpha^2}{\alpha^2 - s^2}\right)^{\lambda} = \left(\frac{1}{1 - \frac{s^2}{\alpha^2}}\right)^{\lambda}, \end{aligned} \tag{5}$$

where K_λ denotes the modified Bessel function. Here we use that $K_\lambda(z) \approx \frac{\Gamma(\lambda)}{2}\left(\frac{2}{z}\right)^\lambda$ for $\lambda > 0$ and $z \downarrow 0$.

Thus, taking $\lambda = 1$ and $\alpha^2 = 2$ in (5), we have that $M_X(s) = \frac{1}{1 - \frac{1}{2}s^2}$, which coincides with (4).

3.1.2 Second Case

Let us consider $\lambda \in \mathbb{N}$, $f_{X|T} = N(0, T)$ and $F_T = GIG(\lambda, 0, \alpha^2) = Ga(\lambda, \alpha^2/2)$ with

$$M_T(s) = \left(1 - \frac{s}{\alpha^2/2}\right)^{-\lambda} = \left(\frac{1}{1 - \frac{s}{\alpha^2/2}}\right)^{\lambda}.$$

Then, the mgf of X is given by

$$M_X(s) = \int_0^\infty M_{X|T}(s) dF_T = \int_0^\infty e^{\frac{1}{2}Ts^2} dF_T = M_T\left(\frac{1}{2}s^2\right) = \left(\frac{1}{1 - \frac{s^2}{\alpha^2}}\right)^{\lambda}, \tag{6}$$

which is a rational function of s. Therefore, in this case $X \sim BME(\boldsymbol{\alpha}_+, \boldsymbol{T}_+, \boldsymbol{t}_+, \boldsymbol{\alpha}_-, \boldsymbol{T}_-, \boldsymbol{t}_-)$, where

$$\boldsymbol{T}_+ = -\boldsymbol{T}_- = \begin{pmatrix} -\alpha & \alpha & 0 & \dots & 0 \\ 0 & -\alpha & \alpha & \dots & 0 \\ \vdots & \vdots & \ddots & \vdots & \\ 0 & 0 & 0 & \dots & -\alpha \end{pmatrix},$$

of dimension $\lambda \times \lambda$, and $\boldsymbol{t}_+ = -\boldsymbol{t}_- = (0, 0, \dots, \alpha)'$ of dimension λ. Depending on λ, the vectors $\boldsymbol{\alpha}_+, \boldsymbol{\alpha}_-$ of dimension λ have special form that can be found using Eq. (2), for example, if $\lambda = 2$ then $\boldsymbol{\alpha}_+ = \boldsymbol{\alpha}_- = (1/4, 1/4)$, if $\lambda = 3$, $\boldsymbol{\alpha}_+ = \boldsymbol{\alpha}_- = (1/8, 3/16, 3/16)$, and so on.

Bayes' theorem implies that $f_{T|X} = GIG(\lambda - 1/2, X^2, \alpha^2)$, and the one-step transition probability function is given by $p(x_t, x_{t+1}) = GH(x_{t+1}; \lambda - 1/2, \alpha, 0, x_t, 0)$, with marginal distribution $GH(\lambda, \alpha, 0, 0, 0)$, whose mgf coincides with (6).

3.1.3 Third Case

Let us consider $f_{X|T} = N(\beta T, T)$, and $F_T = GIG(\lambda, 0, \alpha^2 - \beta^2) = Ga(\lambda, (\alpha^2 - \beta^2)/2)$ where $\lambda \in \mathbb{N}$ (i.e., F_T is an Erlang distribution), thus its mgf is given by

$$M_T(s) = \left(1 - \frac{s}{(\alpha^2 - \beta^2)/2}\right)^{-\lambda} = \left(\frac{1}{1 - \frac{s}{(\alpha^2-\beta^2)/2}}\right)^{\lambda}.$$

Consequently, the mgf of X is given by

$$\begin{aligned} M_X(s) &= \int_0^\infty M_{X|T}(s) dF_T = \int_0^\infty e^{\beta T s + \frac{1}{2} T s^2} dF_T = M_T\left(\beta s + \frac{1}{2} s^2\right) \\ &= \left(\frac{1}{1 - \frac{2\beta s}{\alpha^2-\beta^2} - \frac{s^2}{\alpha^2-\beta^2}}\right)^{\lambda}, \end{aligned} \tag{7}$$

which is a rational function of s, i.e., $X \sim BME(\boldsymbol{\alpha}_+, \boldsymbol{T}_+, \boldsymbol{t}_+, \boldsymbol{\alpha}_-, \boldsymbol{T}_-, \boldsymbol{t}_-)$, where

$$\boldsymbol{T}_+ = \begin{pmatrix} -\alpha+\beta & \alpha-\beta & 0 & \dots & 0 \\ 0 & -\alpha+\beta & \alpha-\beta & \dots & 0 \\ \vdots & \vdots & \ddots & \vdots & \\ 0 & 0 & 0 & \dots & -\alpha+\beta \end{pmatrix}, \quad \boldsymbol{t}_+ = (0, 0, \dots, \alpha - \beta)'$$

$$T_- = \begin{pmatrix} \alpha+\beta & -\alpha-\beta & 0 & \dots & 0 \\ 0 & \alpha+\beta & -\alpha-\beta & \dots & 0 \\ \vdots & \vdots & & \ddots & \vdots \\ 0 & 0 & 0 & \dots & \alpha+\beta \end{pmatrix}, \quad t_- = (0, 0, \dots, -\alpha-\beta)'.$$

The vectors $\boldsymbol{\alpha}_+, \boldsymbol{\alpha}_-$ of dimension λ can be found using Eq. (2).

Bayes' theorem implies that $f_{T|X} = GIG(\lambda - 1/2, X^2, \alpha^2)$, and the one-step probability is given by

$$p(x_t, x_{t+1}) = GH(x_{t+1}; \lambda - 1/2, \sqrt{\alpha^2 + \beta^2}, \beta, x_t, 0), \tag{8}$$

with marginal distribution $GH(\lambda, \alpha, \beta, 0, 0)$, whose mgf coincides with (7).

Having this, and based on [11], we are ready to give the first important definition.

Definition 3.1 A strictly stationary Bilateral Matrix-Exponential ARCH-type model, referred as BME-ARCH(1) model, is a Markov process $\{X_t\}_{t=1}^{\infty}$ with transition distribution given by (8) and marginal distribution BME with moment-generating function given by (7).

3.2 *Multivariate Case*

Now, let us consider the multivariate case, which is a natural extension of the univariate case.

Definition 3.2 [9] A random vector $\boldsymbol{X} \in \mathbb{R}^q$ of dimension q is multivariate Bilateral Matrix-Exponential (MVBME) distributed if the joint moment-generating function $\mathbb{E}(e^{<\boldsymbol{X},\boldsymbol{s}>})$, $\boldsymbol{s} \in \mathbb{R}^q$ is a multidimensional rational function that is a fraction between two multidimensional polynomials. Here $< \cdot, \cdot >$ denotes the inner product in $\mathbb{R}^q$.

In [9], we find the following characterization of this class of distributions.

Theorem 3.1 *A vector* $\boldsymbol{X}$ *follows a multivariate Bilateral Matrix-Exponential distribution if and only if* $< \boldsymbol{X}, \boldsymbol{s} > \sim BME$ *for all* $\boldsymbol{s} \in \mathbb{R}^q/\mathbf{0}$.

Now, consider independent $B_i(t) \sim N(\beta t, t)$, $i = 1, \dots, q$, and

$$f_T = GIG(\lambda, 0, \alpha^2 - \boldsymbol{\beta}'\boldsymbol{\Delta}\boldsymbol{\beta}), \tag{9}$$

where $\lambda \in \mathbb{N}$, $\alpha^2 > \boldsymbol{\beta}'\boldsymbol{\Delta}\boldsymbol{\beta}$, $\boldsymbol{\Delta} \in \mathbb{R}^{q\times q}$ is a positive definite matrix with $|\boldsymbol{\Delta}| = 1$, and $\boldsymbol{\beta} = (\beta, \dots, \beta)' \in \mathbb{R}^q$, i.e., $\boldsymbol{\beta}$ is a vector of dimension $q \times 1$.

Equation (9) denotes the density of a Gamma distribution (in particular an Erlang) with parameters $(\lambda, (\alpha^2 - \boldsymbol{\beta}'\boldsymbol{\Delta}\boldsymbol{\beta})/2)$, i.e., we have that its mgf is given by

$$M_T(s) = \left(1 - \frac{s}{(\alpha^2 - \boldsymbol{\beta}'\boldsymbol{\Delta}\boldsymbol{\beta})/2}\right)^{-\lambda} = \left(\frac{1}{1 - \frac{s}{(\alpha^2 - \boldsymbol{\beta}'\boldsymbol{\Delta}\boldsymbol{\beta})/2}}\right)^{\lambda}.$$

Define $\mathbf{X}|T = (B_1(T), \ldots, B_q(T))'$, i.e., $f_{\mathbf{X}|T} = N_q(T\boldsymbol{\Delta}\boldsymbol{\beta}, T\boldsymbol{\Delta})$. Thus, the mgf of $\boldsymbol{X}$ is given by

$$\begin{aligned} M_{\mathbf{X}}(\boldsymbol{s}) &= \int_0^\infty M_{X|T}(s)dF_T = \int_0^\infty \exp\left(\boldsymbol{s}'T\boldsymbol{\Delta}\boldsymbol{\beta} + \frac{1}{2}\boldsymbol{s}'T\boldsymbol{\Delta}\boldsymbol{s}\right)dF_T \\ &= \int_0^\infty \exp(\theta T)dF_T; \quad \text{where } \theta = \boldsymbol{s}'\boldsymbol{\Delta}\boldsymbol{\beta} + \frac{1}{2}\boldsymbol{s}'\boldsymbol{\Delta}\boldsymbol{s} \\ &= M_T(\theta) = \left(\frac{1}{1 - \frac{\theta}{(\alpha^2 - \boldsymbol{\beta}'\boldsymbol{\Delta}\boldsymbol{\beta})/2}}\right)^{\lambda} = \left(\frac{1}{1 - \frac{2\boldsymbol{s}'\boldsymbol{\Delta}\boldsymbol{\beta}}{\alpha^2 - \boldsymbol{\beta}'\boldsymbol{\Delta}\boldsymbol{\beta}} - \frac{\boldsymbol{s}'\boldsymbol{\Delta}\boldsymbol{s}}{\alpha^2 - \boldsymbol{\beta}'\boldsymbol{\Delta}\boldsymbol{\beta}}}\right)^{\lambda} \end{aligned} \tag{10}$$

which is a rational function of $\boldsymbol{s}$. This implies that $\boldsymbol{X} \sim MVBME$, and thus $< \boldsymbol{X}, \boldsymbol{s} > \sim BME$ (see Sect. 3.1.3).

On the other hand, Bayes' theorem implies

$$f_{T|\mathbf{X}} = GIG\left(\lambda - \frac{q}{2}, r^2, \alpha^2\right)$$

where $r = \sqrt{\mathbf{x}'\boldsymbol{\Delta}^{-1}\mathbf{x}}$ with $\mathbf{x} \in \mathbb{R}^q$.

Following similar steps to the univariate case, we can construct an i-order Markov transition probability function as follows:

$$\begin{aligned} p(x_{t+i}|\mathbf{x}^{(t,i-1)}) &= \int_{\mathbb{R}^+} N(x_{t+i}; T\beta_{t+i}, T)GIG\left(T; \lambda - \frac{i}{2}, r^2_{(t,i-1)}, \alpha^2\right)dT \\ &= GH\left(x_{t+i}; \lambda - \frac{i}{2}, \sqrt{\alpha^2 + \beta^2_{t+i}}, \beta_{t+i}, r_{(t,i-1)}, 0\right) \end{aligned} \tag{11}$$

where $r_{(t,i-1)} = \sqrt{\mathbf{x}^{(t,i-1)\prime}\boldsymbol{\Delta}^{-1}\mathbf{x}^{(t,i-1)}}$, and $\mathbf{x}^{(t,i-1)}$ is an i-dimensional vector denoting the time-space values corresponding to $\boldsymbol{X}^{(t,i-1)} = (X_t, \ldots, X_{t+i-1})$.

According to the previous section, the marginal distribution is given by $GH(\lambda, \alpha, \beta, 0, 0)$ with mgf on the form (7), which belongs to the BME class.

Now, we are ready to address the second important definition of this paper.

Definition 3.3 Suppose that $\{X_t\}_{t\geq 1}$ has a marginal distribution $GH(\lambda, \alpha, \beta, 0, 0)$, and $X_{t+i}|\mathbf{X}^{(t,i-1)}$, $i = 1, \ldots, q$, follows the distribution given in (11) (so that marginally $X_{t+q} \sim GH(\lambda, \alpha, \beta, 0, 0)$). The resulting Markov process $\{X_t\}_{t\geq 1}$ will be termed as the stationary BME-ARCH (q) model.

4 Construction Using Matrix-Exponential Distributions

From [2, 3], another characterization of the Matrix-Exponential (ME) distributions is that they are the absolutely continuous distributions on the positive real line with densities f that are trigonometric polynomials:

$$Z \sim ME, \quad \text{with} \quad f_Z(z) = \sum_{j=0}^{q} c_j z^{n_j} e^{\theta_j z},$$

where q and the n_j's are nonnegative integers, and the c_j's and θ_j's are complex.

For the sake of notation, we denote $\eta_j = -\theta_j$, and suppose $\eta_j > 0$. Taking $f_{X|Z} = N(0, Z)$ (see [9]), we have that the marginal of X is given by

$$\begin{aligned}
f_X(x) &= \int_0^\infty f_{X|Z}(x|z) \cdot f_Z(z)dz = \int_0^\infty \frac{1}{\sqrt{2\pi}} z^{-1/2} \exp\left(-\frac{1}{2}x^2 z^{-1}\right) \sum_{j=0}^{q} c_j z^{n_j} e^{-\eta_j z} dz \\
&= \sum_{j=0}^{q} c_j \frac{1}{\sqrt{2\pi}} \int_0^\infty (z^{-1/2} z^{n_j}) \exp\left(-\frac{1}{2}x^2 z^{-1} - \eta_j z\right) dz \\
&= \sum_{j=0}^{q} c_j \frac{1}{\sqrt{2\pi}} \int_0^\infty z^{\lambda_j - 1} \exp\left(-\frac{1}{2}(x^2 z^{-1} + 2\eta_j z)\right) dz; \quad \lambda_j - 1 = -1/2 + n_j \\
&= \sum_{j=0}^{q} c_j \frac{1}{\sqrt{2\pi}} \frac{2x^{\lambda_j} K_{\lambda_j}(\alpha x)}{\alpha^{\lambda_j}}; \quad \alpha^2 = 2\eta_j \\
&= \sum_{j=0}^{q} \left[c_j \frac{1}{\sqrt{2\pi}} \frac{2}{\alpha^{\lambda_j}} \frac{1}{a(\lambda, \alpha, 0, 0)} \right] \left[a(\lambda, \alpha, 0, 0) x^{\lambda - 1/2} K_{\lambda - 1/2}(\alpha x) \right]; \quad \lambda = \lambda_j + 1/2 \\
&= \sum_{j=0}^{q} c_j^* GH(x; \lambda, \alpha, 0, 0, 0) \quad \text{(note that both } \lambda \text{ and } \alpha \text{ depend on } j\text{)}
\end{aligned}$$

where

$$c_j^* = c_j \frac{1}{\sqrt{2\pi}} \frac{2}{\alpha^{\lambda_j}} \frac{1}{a(\lambda, \alpha, 0, 0)}, \quad \text{and} \quad a(\lambda, \alpha, 0, 0) = \frac{\alpha^{\lambda + 1/2}}{\sqrt{2\pi}\, \Gamma(\lambda) 2^{\lambda - 1}}.$$

Thus

$$c_j^* = c_j \Gamma(\lambda) \left(\frac{2}{\alpha^2}\right)^\lambda.$$

The mgf of X is given by

$$M_X(s) = \sum_j c_j^* \left(\frac{1}{1 - s^2/\alpha^2}\right)^\lambda.$$

Note this is a rational function since λ is an integer ($\lambda = 1 + n_j$). Thus, the marginal is a BME distribution (see Sect. 3.1.2).

Now, the conditional density function can be obtained as

$$
\begin{aligned}
f_{Z|X}(z|x) &\propto f_{X|Z}(x|z) \cdot f_Z(z) \propto z^{-1/2} \exp\left(-\frac{1}{2}x^2 z^{-1}\right) \sum_{j=0}^{q} c_j z^{n_j} e^{-\eta_j z} \\
&= \sum_{j=0}^{q} c_j \left(z^{-1/2} z^{n_j}\right) \exp\left(-\frac{1}{2}x^2 z^{-1} - \eta_j z\right) \\
&= \sum_{j=0}^{q} c_j z^{\lambda_j - 1} \exp\left(-\frac{1}{2}(x^2 z^{-1} + 2\eta_j z)\right); \quad \text{where } \lambda_j - 1 = -1/2 + n_j \\
&\propto \sum_{j=0}^{q} c_j GIG(z; \lambda_j, x^2, 2\eta_j).
\end{aligned}
$$

The one-step transitions are given by

$$
\begin{aligned}
p(x_t, x_{t+1}) &= \int_0^\infty f_{X_{t+1}|Z} \cdot f_{Z|X_t} dz \\
&= \int_0^\infty N(x_{t+1}; 0, z) \sum_j c_j GIG(z; \lambda_j, x_t^2, 2\eta_j) dz \\
&= \sum_j c_j \left(\int_0^\infty N(x_{t+1}; 0, z) GIG(z; \lambda_j, x_t^2, 2\eta_j) dz\right) \\
&= \sum_j c_j GH(x_{t+1}; \lambda_j, \sqrt{2\eta_j}, 0, x_t, 0).
\end{aligned}
$$

With marginal distribution

$$
\sum_j c_j^* GH(\lambda_j + 1/2, \sqrt{2\eta_j}, 0, 0, 0)
$$

which is a BME distribution (see Sect. 3.1.2).

Example 4.1 Hyper-exponential distribution

Consider k random variables $Y_i \sim \exp(\lambda_i)$, $i = 1, 2, \ldots, k$, and assume that Z takes the value of Y_i with probability γ_i. The distribution of Z, called hyper-exponential distribution, can be expressed as a proper mixture of the Y_i's. A ME representation is given by

$$\boldsymbol{\gamma} = (\gamma_1, \ldots, \gamma_k), \quad \boldsymbol{L} = \begin{pmatrix} -\lambda_1 & 0 & \ldots & 0 \\ 0 & -\lambda_2 & \ldots & 0 \\ \vdots & \vdots & \ddots & \vdots \\ 0 & 0 & \ldots & -\lambda_k \end{pmatrix}, \quad \boldsymbol{\ell} = (\lambda_1, \lambda_2, \ldots, \lambda_k)',$$

with density function given by $f_Z(z) = \sum_{i=1}^k \gamma_i \lambda_i e^{-\lambda_i z}$.

Taking the conditional density function as $f_{X|Z} = N(0, Z)$ then

$$f_{Z|X} = \sum_{i=1}^{k} \gamma_i \lambda_i GIG(z; 1/2, x^2, 2\lambda_i),$$

and the one-step transition probability is given by

$$p(x_t, x_{t+1}) = \sum_{i=1}^{k} \gamma_i \lambda_i GH(x_{t+1}; 1/2, \sqrt{2\lambda_i}, 0, x_t, 0),$$

with marginal distribution $\sum_{i=1}^k \gamma_i GH(1, \sqrt{2\lambda_i}, 0, 0, 0)$, and mgf $\sum_{i=1}^k \gamma_i \left(\frac{1}{1-s^2/2\lambda_i}\right)$.

Indeed, we can also compute some useful moments.

1. The first moment: $\mathbb{E}(X) = \int_0^\infty \mathbb{E}(X|Z) dF_Z$. If $f_{X|Z} = N(0, Z)$, then $\mathbb{E}(X|Z) = 0$, i.e., $\mathbb{E}(X) = 0$.
2. Second moment: $\mathbb{E}(X^2) = \int_0^\infty \mathbb{E}(X^2|Z) dF_Z = \int_0^\infty z dF_Z = \mathbb{E}(Z)$. We know that if $Z \sim ME(\boldsymbol{\gamma}, \boldsymbol{L}, \boldsymbol{\ell})$, its moments are given by $M_i = \mathbb{E}(Z^i) = i!\boldsymbol{\gamma}(-\boldsymbol{L})^{-(i+1)}\boldsymbol{\ell}$. In particular, for Example 4.1, we get that $\mathbb{E}(X^2) = \mathbb{E}(Z) = \sum_{i=1}^k \frac{\gamma_i}{\lambda_i}$.
3. Third moment: $\mathbb{E}(X^3) = \int_0^\infty \mathbb{E}(X^3|Z) dF_Z = 0$.
4. Fourth moment: $\mathbb{E}(X^4) = \int_0^\infty \mathbb{E}(X^4|Z) dF_Z = \int_0^\infty 3z^2 dF_Z = 3\mathbb{E}(Z^2)$. Since $Z \sim$ Hyper-exponential then

$$\mathbb{E}(X^4) = 3\sum_{i=1}^{k} \frac{2\gamma_i}{\lambda_i^2} = 6\sum_{i=1}^{k} \frac{\gamma_i}{\lambda_i^2}.$$

Note that we could also use Proposition 1 from [11] in order to find the moments.

Now, we will present the third important result from this paper: the construction of a BME-normal model.

4.1 A BME-Normal Model

It is well known that for a nonnegative random variable $Z \sim ME(\boldsymbol{\gamma}, \boldsymbol{L}, \boldsymbol{\ell})$, with $\boldsymbol{\ell} = -\boldsymbol{L}\boldsymbol{e}$, its mgf is given by

$$M_Z(s) = \boldsymbol{\gamma}(s\boldsymbol{L}^{-1} + \boldsymbol{I})^{-1}\boldsymbol{e},$$

where $\boldsymbol{e}$ is a vector of appropriate dimension of ones.

The construction of the BME-normal model is as follows.

Let $Z \sim ME(\boldsymbol{\gamma}, \boldsymbol{L}, \boldsymbol{\ell})$, $\tau > 0$, and consider $X|Z \sim N(Z\mu, Z\tau)$. Then the mgf of X is given by

$$M_X(s) = \int_0^\infty \exp\left(zs\mu + z\frac{1}{2}s^2\tau\right) dF(z) = \boldsymbol{\gamma}(\theta\boldsymbol{L}^{-1} + \boldsymbol{I})^{-1}\boldsymbol{e}$$

where $\theta = s\mu + \frac{1}{2}s^2\tau$. Since this is a rational function of s, thus $X \sim BME$, whose matrix representation depends on $\boldsymbol{\gamma}$ and $\boldsymbol{L}$.

Imposing $Y|(X, Z) \sim N\left(\frac{X}{Z}, \frac{\phi\tau}{Z}\right)$, with $\phi > 0$, then

$$X|(Y, Z) \sim N\left(Z\frac{Y + \phi\mu}{1 + \phi}, Z\frac{\phi\tau}{1 + \phi}\right).$$

The n-steps transition probability can be obtained from

$$X_n|(X_0, Z) \sim N(a_n X_0 + Zb_n, Zc_n)$$

where

$$\begin{aligned}
a_n &= a_1^n = \frac{1}{(1 + \phi)^n} \\
b_n &= b_1(1 + a_1 + \cdots + a_1^{n-1}) = \mu(1 - (1 + \phi)^{-n}) \\
c_n &= c_1(1 + a_1^2 + \cdots + a_1^{2(n-1)}) = \tau(1 - (1 + \phi)^{-2n}).
\end{aligned}$$

See [12] for more details.

The asymptotic distribution can be obtained taking $n \to \infty$, thus $a_n \to 0$, $b_n \to \mu$, and $c_n \to \tau$, i.e., $X_\infty|Z \sim N(Z\mu, Z\tau)$, and $X_\infty \sim BME$.

Note that we can write

$$X_n|X_{n-1} = AX_{n-1} + B + C\varepsilon_n,$$

where

$$A = \frac{1}{1 + \phi}, \quad B = Z\frac{\phi\mu}{1 + \phi}, \quad C = (Z\tau(1 - (1 + \phi)^{-2}))^{1/2}, \quad \varepsilon_n \sim N(0, 1).$$

The process $\{X_n\}_{n=1}^\infty$ is stationary.

A simulation algorithm for this process is the following:

Algorithm 1 Simulation of $\{X_n\}_{n\geq 1}$

For $\phi > 0$.

1: Simulate $Z \sim ME(\boldsymbol{\gamma}, \boldsymbol{L}, \boldsymbol{\ell})$
2: Simulate $X_0 \sim N(Z\mu, Z\tau)$
3: Generate $\varepsilon_1, \varepsilon_2, \cdots \sim N(0, 1)$
4: $X_1 = \frac{1}{1+\phi} X_0 + Z \frac{\phi\mu}{1+\phi} + (Z\tau(1 - (1+\phi)^{-2}))^{1/2} \varepsilon_1$
5: $X_2 = \frac{1}{1+\phi} X_1 + Z \frac{\phi\mu}{1+\phi} + (Z\tau(1 - (1+\phi)^{-2}))^{1/2} \varepsilon_2$

$\vdots$

An example of the simulation of this process is the following.

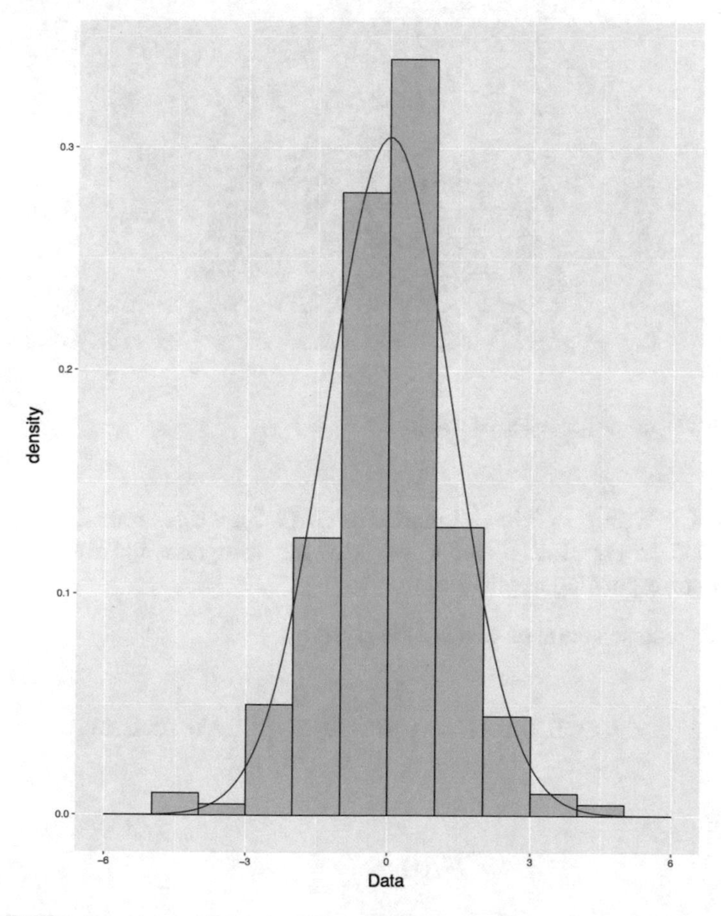

Fig. 1 All 200 trajectories and fixed size 900

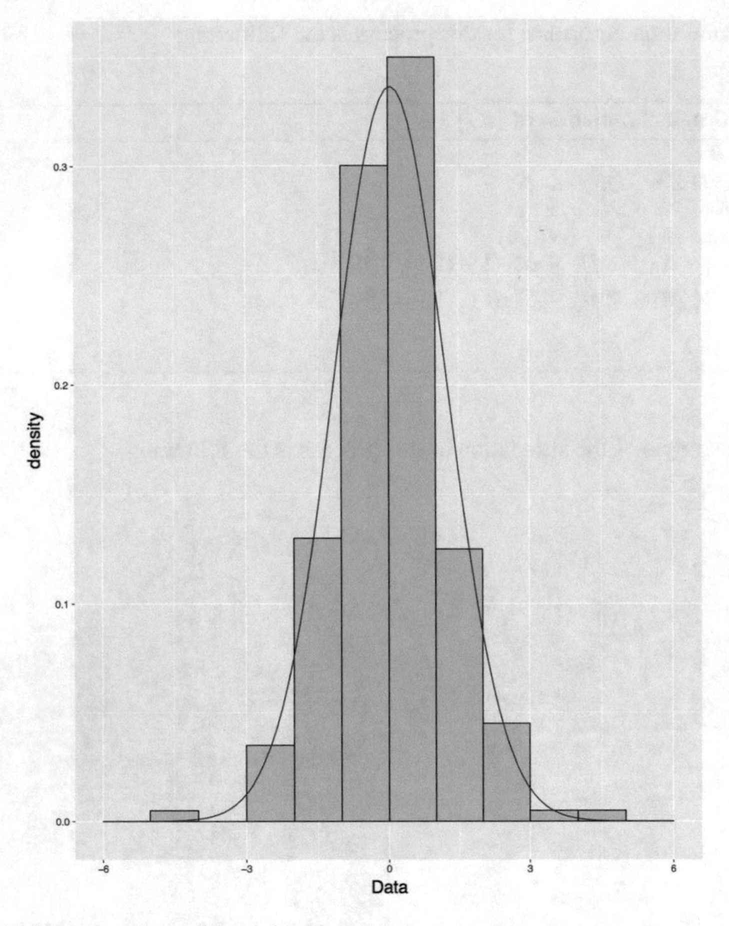

Fig. 2 All 200 trajectories and fixed size 10

Example 4.2 With $\mu = 0$, $\tau = 1$, and $Z \sim Ga(3, 2)$, we generate 200 trajectories of size 1000. In Figs. 1, 2, 3, and 4, we show the histograms for different samples and their corresponding normal fitting.

An ME representation of $Ga(3, 2)$ is given by

$$\boldsymbol{\gamma} = (1, 0, 0); \quad \boldsymbol{L} = \begin{pmatrix} -2 & 2 & 0 \\ 0 & -2 & 2 \\ 0 & 0 & -2 \end{pmatrix}; \quad \boldsymbol{\ell} = (0, 0, 2)',$$

with mgf

$$M_z(s) = \frac{1}{\left(1 - \frac{s}{2}\right)^3}.$$

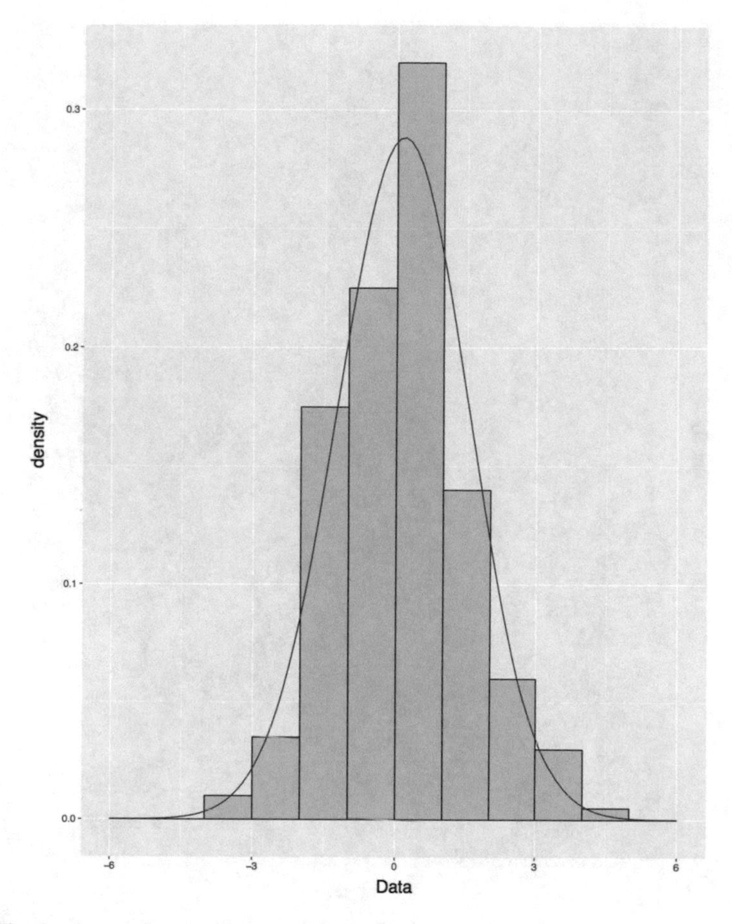

Fig. 3 Fixed trajectory 1 and sizes $n = 801, \ldots, 1000$

Since we can write $X_0 = ZX_0^*$ with $X_0^*|Z \sim N(\mu, \tau/Z)$, then, as a particular case of the models presented in [12] in continuous time, we have that

$$X_t|(X_0^*, Z) \sim N(Z(a_t X_0^* + b_t), Zc_t)$$

where $a_t = (1+\phi)^{-t}$; $\quad b_t = \mu(1-(1+\phi)^{-t})$; $\quad c_t = \tau(1-(1+\phi)^{-2t})$, for $\phi > 0$ and $t > 0$.

The Chapman–Kolmogorov (CK) equation is satisfied if $\mathbb{E}(M_{X_{t+s}|X_s}(\lambda)) = M_{X_t|X_0}(\lambda)$. Since

$$M_{X_t|X_0}(\lambda) = \boldsymbol{\gamma}(\theta_t \boldsymbol{L}^{-1} + \boldsymbol{I})^{-1}\boldsymbol{e}$$

where $\theta_t = \lambda(a_t x_0^* + b_t) + \frac{1}{2}\lambda^2 c_t$, we get that

$$\mathbb{E}(M_{X_{t+s}|X_s}(\lambda)) = \boldsymbol{\gamma}(\theta_{t+s}\boldsymbol{L}^{-1} + \boldsymbol{I})^{-1}\boldsymbol{e}$$

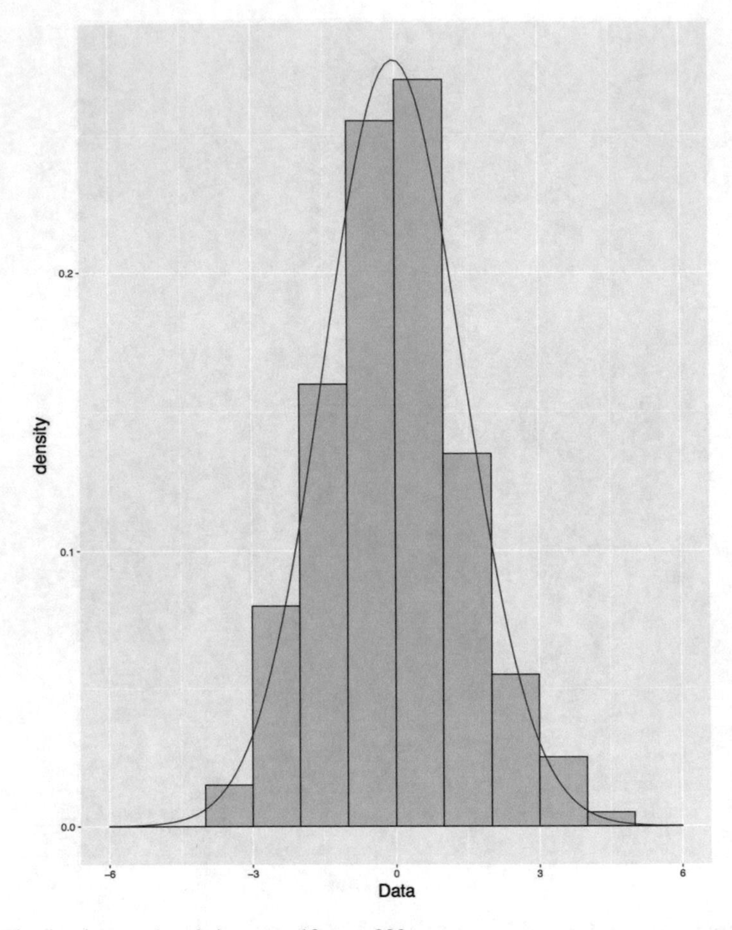

Fig. 4 Fixed trajectory 1 and sizes $n = 10, \ldots, 209$

where $\theta_{t+s} = \lambda(a_{t+s}x_0^* + b_{t+s}) + \frac{1}{2}\lambda^2 c_{t+s}$. (See [12] for more details). Thus, CK is satisfied.

5 Conclusions

In this paper, we have considered a wide class of distributions called Matrix-Exponential (ME) which is dense in $\mathbb{R}_+^d$, and it is a robust family of distributions. In order to extend this class into the real line, we have the Bilateral Matrix-Exponential (BME) distributions, which are defined as the random variables with rational moment-generating function.

In the area of stochastic models, the classes of ME and BME are well known, although in Bayes community they are not. Some Markov Chain Monte Carlo algorithms, like the Metropolis–Hastings and the Gibbs sampler, have been used in order to estimate a sub-class of the ME distributions called phase-type distributions. Exploring the construction of stationary processes using the ME and BME classes of distributions represents an innovative way of modeling.

Indeed, we have provided interesting alternatives on constructing stationary Markov processes having BME as marginal distributions, using their characterization of having rational moment-generating functions. As an extension, we can also use the BME marginals considering the matrix-analytic representation, even considering rewards, in order to extend their field of applications.

Acknowledgements The author gratefully acknowledges the support of a CONACyT postdoctoral fellowship at IIMAS that gave origin to the present work.

References

1. Ahn, S., Ramaswami, V.: Bilateral Phase-type distributions. Stoch. Model. **21**, 239–259 (2005)
2. Asmussen, S., Bladt, M.: Renewal theory and queueing algorithms for Matrix-Exponential distributions. Matrix-Analytic Methods in Stochastic Models, pp. 313–341 (1997)
3. Asmussen, S., O' Cinneide, C.A.: Matrix-Exponential Distributions. Encyclopedia of Statistical Sciences. Wiley Online Library, Hoboken (1998)
4. Barndorff-Nielsen, O.E.: Exponentially decreasing distributions for the logarithm of particle size. Proc. R. Soc. Lond. **353**, 401–419 (1977)
5. Bean, N.G., Nielsen, B.F.: Quasi-Birth-and-Death process with rational arrival process components. Stoch. Model. **26**, 309–334 (2010)
6. Bladt, M., Nielsen, B.F.: Multivariate matrix-exponential distributions. Stoch. Model. **1**(26), 1–26 (2010)
7. Bladt, M., Nielsen, B.F.: On the representation of distributions with rational moment generating functions. Kgs. Lyngby: Technical University of Denmark (DTU). (DTU Compute. Technical Report; No. 2012–16) (2012)
8. Bladt, M., Neuts, M.F.: Matrix-exponential distributions: calculus and interpretations via flows. Stoch. Model. **19**, 113–124 (2003)
9. Bladt M., Esparza L.J.R., Nielsen B.F.: Bilateral Matrix-Exponential Distributions. In: Latouche G., Ramaswami V., Sethuraman J., Sigman K., Squillante M., Yao D. (eds.) Matrix-Analytic Methods in Stochastic Models. Springer Proceedings in Mathematics & Statistics, vol. 27. Springer, New York, NY (2013)
10. Kulkarni, V.G.: A new class of multivariate phase-type distributions. Oper. Res. **37**, 151–158 (1989)
11. Mena, R.H., Walker, S.G.: On the stationary version of the generalized hyperbolic ARCH model. AISM **59**, 325–348 (2007)
12. Mena, R.H., Walker, S.G.: On a construction of Markov models in continuous time. Int. J. Stat. **3**(LXVII), 303–323 (2009)
13. Neuts, M.F.: Probability distributions of phase-type. In: Liber Amicorum Prof. Emeritus H. Florin, pp. 173–206 (1975)
14. Neuts, M.F.: Matrix Geometric Solutions in Stochastic Models, vol. 2. Johns Hopkins University Press, Baltimore, MD (1981)
15. Pitt, M., Chatfield, C., Walker, S.G.: Constructing first order autoregressive models via latent processes. Scand. J. Stat. **29**, 657–663 (2002)

BoostNet: Bootstrapping Detection of Socialbots, and a Case Study from Guatemala

E. I. Velazquez Richards, E. Gallagher and P. Suárez-Serrato

Abstract We present a method to reconstruct networks of socialbots given minimal input. Then we use Kernel Density Estimates of Botometer scores from 47,000 social networking accounts to find clusters of automated accounts, discovering over 5,000 socialbots. This statistical and data-driven approach allows for inference of thresholds for socialbot detection, as illustrated in a case study we present from Guatemala.

Keywords Kernel decomposition estimate · Data analysis · Social network analysis · Empirical data

1 Introduction

In this paper, we analyze data from the social networking platform Twitter. We use a statistical approach, with bi-variate Kernel Density Estimates, to detect automated accounts (socialbots) at scale in a large dataset. We present our BoostNet algorithm, which allows for the detection of networks of socialbots in microblogs and social media platforms given a very small number of initial accounts. We illustrate its performance with empirical data collected from Twitter in relation to current events in Guatemala.

To begin to describe some of the context of the events that have led to this particular social media situation, first we point out that the displacement of people due to armed conflict and corruption is a problem that affects many countries around the world. This phenomenon has strongly affected the Central American countries of

E. I. Velazquez Richards · P. Suárez-Serrato (✉)
Instituto de Matemáticas, Universidad Nacional Autónoma de México, Ciudad Universitaria, 04510 Coyoacán, Mexico City, Mexico
e-mail: pablo@im.unam.mx

E. Gallagher
Integrative Media, Pursuance Project, 3419 Westminster Avenue #25, Dallas, TX 75205, USA

P. Suárez-Serrato
Department of Mathematics, University of California Santa Barbara, Goleta, CA, USA

I. Antoniano-Villalobos et al. (eds.), *Selected Contributions on Statistics and Data Science in Latin America*, Springer Proceedings in Mathematics & Statistics 301,
https://doi.org/10.1007/978-3-030-31551-1_11

Honduras and Guatemala. Nevertheless, currently the US enjoys the lowest level of undocumented immigrants in US in a decade, according to a Pew Research Center analysis of government data [12]. The same study indicates that border apprehensions have declined for Mexicans but risen for other Central Americans.

What are the root causes of migration? Understanding these can better help prevent forced displacement of people and thus also the effects on societies that receive them. In a previous work, we investigated the use of socialbots in Honduras in relation to protests alleging electoral fraud [9].

Consider the case of Guatemala. The International Commission against Impunity in Guatemala (CICIG https://www.cicig.org/ [4]) was created in 2006 by the United Nations and Guatemala. It is an international body whose mission is to investigate and prosecute serious crime [4].

An independent international body, CICIG investigates illegal security groups and clandestine security organizations in Guatemala. These are criminal groups believed to have infiltrated state institutions, fostering impunity, and undermining democratic advances since the end of the armed conflict in the 1990s. The third impeachment against President Jimmy Morales for illicit electoral financing during his electoral campaign in 2015 was requested by the Attorney General and the CICIG.

The mandate of the CICIG was set to end originally on September 3, 2019, but it has been cut abruptly short as Guatemalan President Morales ordered the CICIG to leave the country on January 7, 2019 [11].

After we published our work on socialbots in Honduras [9], we were contacted by a Guatemalan journalist claiming that similar socialbots were acting against the population there. It was claimed that multiple Twitter accounts were being used to systematically intimidate and harass members of the CICIG and the media that covers their activities. In April 2018, we were provided with 19 seed accounts of potential socialbots that were notorious in this instance for their negative behavior.

From these 19 accounts, we reconstructed a network of over 35,000 accounts, by collecting their followers and their followees. The rationale is that socialbot accounts are not generally followed by human accounts. Following this premise, we begin with these 19 seed accounts and take two hops out into the follower network to find potential accounts that are also automated and being used for this purpose. This method, which we call BoostNet, is explained in Algorithm 1 and the networks are visualized in Fig. 1 in terms of reach and spread of the full network, and in Fig. 2 in a subset of the most active bot accounts and their retweet relationships. This strategy led us to discover a socialbot network of over 3000 accounts. To this end, we queried *Botometer* [6] and performed a statistical analysis of the scores it provides to find the network of socialbots (explained below, see Fig. 3). Botometer is a machine learning algorithm that is trained to categorize Twitter accounts as human or bot based on a list of tens of thousands of previously labeled examples. The algorithm scans account for about 1,200 different characteristics and behavior patterns and ranks the account on a scale of probability whether the account is automated or human-operated.

We further validated our method by using 14 more accounts mentioned in a media interview about socialbot harassment in Guatemala from November 2018. From these 14 seed accounts, we reconstructed a full network of over 12,000 accounts and found

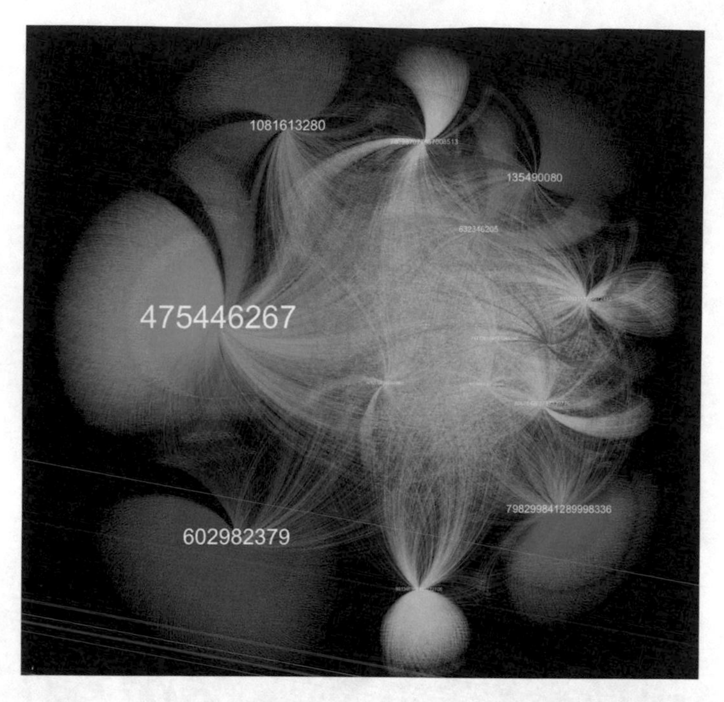

Fig. 1 Gephi network graph created using OpenOrd and Force Atlas 2 force-directed layout algorithms. The network contains 35,208 nodes, 59,471 edges, and 8 distinct clusters or communities. As per Twitter's data policies, we have used the user ID to label the nodes, and not the account handle

over 2,000 socialbots (see Fig. 4). There were over 600 socialbots common to both datasets.

In order to better understand the magnitude of this socialbot network, it is helpful to observe that Guatemala has a population of around 17 million people, and Internet users include only 4.5 million [10]. Measurements of social media use in Guatemala indicate that 5.24% of Internet users are active on Twitter [13]. We can therefore extrapolate an—admittedly rough—estimate of around 250,000 Twitter users in Guatemala (2018 figures). In this perspective, socialbot networks of 3,000 and 2,000 accounts can have a considerable impact.

Finally, at the request of a reviewer, we also include the use of a novel method to quantify the topological effect that bot accounts create in a network. We use a new graph distance [16] that recovers topological features found via non-backtracking cycles—and is inspired by rigidity properties of the marked length spectrum—to precisely measure how a network is affected by the presence of bots. These results are summarized in Table 2.

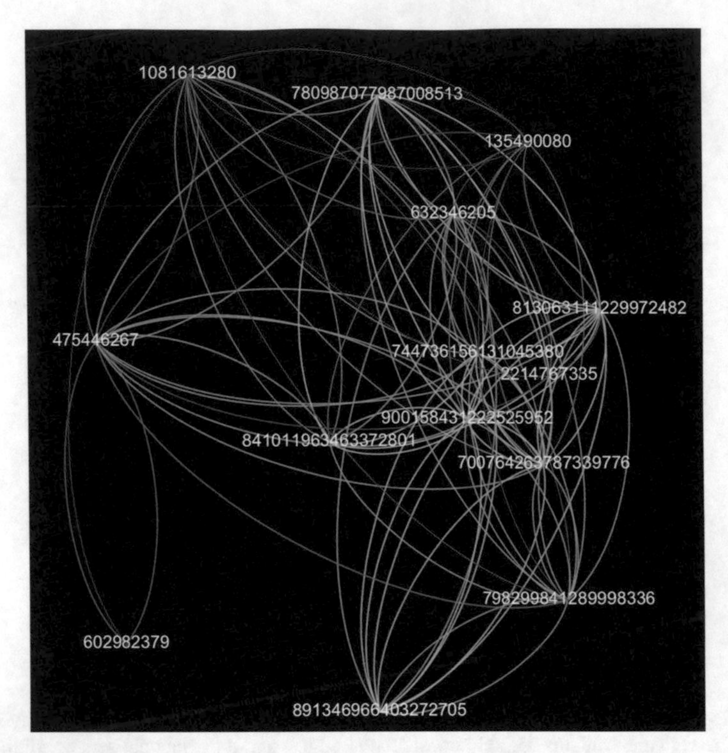

Fig. 2 Gephi network graph created using OpenOrd and Force Atlas 2 force-directed layout algorithms. The complete network (see Fig. 1) contains 35,208 nodes, 59,471 edges, and 8 distinct clusters or communities, which were filtered by degree range 50 revealing 14 visible nodes (0.04%) and 100 visible edges (0.17%) of the complete network

2 Discussion

The influence of socialbots in political life can restrict free speech and disband activist groups. We have researched the effect of socialbots previously in case studies on Mexico and Honduras. A study by Freedom House now includes pro-government pundits and bots to their considerations of government censorship online as they are capable of altering political dialogue.

Fake Twitter users among the followers of political figures are now a common phenomenon, known as *astro-turf*, with up to 20–29% fake followers in some prominent cases. Keeping track of socialbots is therefore important, for example, in electoral races and in the recognition of *influencers*.

To evaluate the role of bots, it is also possible to cluster opinions using hashtags. These methods depend on accessibility of data. It can become expensive to buy a complete dataset, and the sampling method of Twitter's Stream API introduces additional statistical uncertainties. Here, we circumvent these issues by focusing on an initial set of accounts and reconstruct a network of linked accounts.

Post sentiment has been shown to be efficient at separating human from nonhuman users in Twitter; however, these methods work best in English. This is why in this paper we concentrate on nonlanguage-specific features, as the language processing applications to Spanish—as well as for other low resource languages—are not yet available.

3 Data Collection

In this section, we describe our strategy to gather a large network of socialbot accounts from a small number of accounts that are reported to be abusing a social media service. We present an algorithm that can be replicated in other circumstances, and can be easily implemented to reconstruct a complete network of linked accounts.

3.1 BoostNet: A Method to Find Socialbot Networks with Minimal Input

The following pseudocode illustrates our workflow to construct networks where the human and socialbot accounts can be analyzed. Our method allows us to find large networks of socialbots given a small number of starting accounts. We illustrate its performance with an empirical case study here, we discovered two sets of socialbots: one containing over 3,000 socialbot accounts and the second containing over 2,000 socialbot accounts, starting from only 19 and 14 accounts, respectively, in each case that were reportedly harassing journalists and members of the CICIG.

Algorithm 1 BoostNet : Bootstrapping Socialbot Network Detection

Require: A collection C of Twitter Accounts
Ensure: Full linked network $N(C)$ of with Socialbot account score
1: **Initialize**
2: For each account a in C:
3: Collect followers $F(C)$ of the collection C from Twitter's Rest API
4: Collect those accounts $FR(a)$ who are following a from Twitter's Rest API
5: Obtain scores of every account in $F(C)$, $FR(a)$ and a from Botometer, to determine if it is Human or Socialbot
6: Construct a follower–followee network $N(C)$ annotated with Botometer scores
7: **return** $N(C)$
8: **End**

3.2 Comparison with Twitter's Stream API

One poignant criticism of certain Twitter studies is the reliance on Twitter's Streaming API for data acquisition. While Twitter's Streaming API provides free and public access to a sample of tweets and has promoted research into social networks, there are certain limitations that its sampling method impose. Here we circumvent these difficulties in finding networks of linked accounts. Connections of followers and followees were queried from Twitter's Rest API. In this way, we have reconstructed a full dataset of accounts that are linked in the same connected network.

Certain studies have avoided this sampling bias uncertainty from Twitter's Streaming API by using the Search API to obtain complete datasets [14]. Another option seems to work directly with Twitter, and some research has been successful at establishing influence relations using this kind of access [1].

For this work, we have reconstructed a full dataset of interest for our research using the Rest API only.

4 Statistical Detection of Socialbot Networks

For a review of *Botometer*, we recommend [8]. Socialbots have been employed for political purposes [20]. It has also been observed that this technology is used in marketing and propaganda [18]. Although research has uncovered other successful methods of bot detection [2, 3, 5, 7], *Botometer* provides public API access. The features it has built in as well as a review of how it compares to, and surpasses, other methods can be found in [6, 17].

We have previously used this method for identifying bots in online communities in Latin America, specifically in Mexico and Honduras [9, 15, 19].

In this work, we have concentrated on three of the nonlanguage-specific classifiers that Botometer provides. Botometer is a supervised learning tool that detects bot accounts. It exploits over 1,200 features, including user metadata, social contacts, diffusion networks, content, sentiment, and temporal patterns. Evaluated on a large dataset of labeled accounts, it has reported high accuracy in discerning human from bot accounts.

Network features capture various dimensions of information diffusion patterns. User features are based on Twitter metadata related to an account, including language geographic locations and account creation time. Friend features include descriptive statistics relative to account contacts such as median moments and entropy of the distribution of the followers, followees, and posts. Temporal features capture timing patterns of content generation and retweets, for example, how that signal is similar to a specific process (Poisson) or the average time between two consecutive posts.

Using the scores from Temporal, Network, and Friend evaluations that each account in our dataset yields, we aggregate this data and then find a 2D bi-modal behavior using KDE, as illustrated in Figs. 3 and 4.

It is important to highlight that a nonparametric method is needed in order to infer the possible separation between types of accounts (human or bot) without making any a priori assumptions about how these are distributed or the mass that should be associated to them. Using kernel density estimates allows us to visualize different

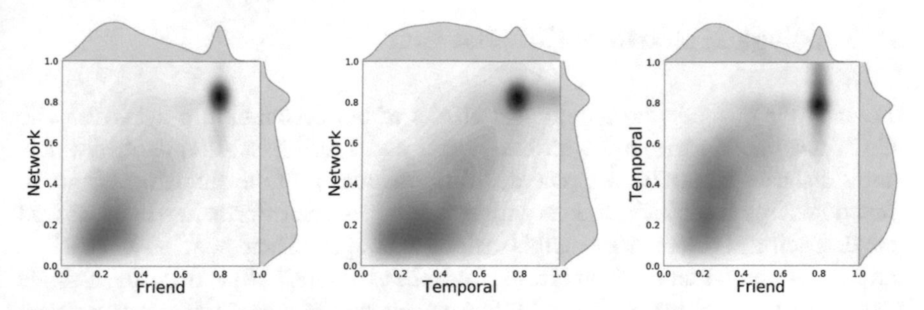

Fig. 3 2D Kernel decomposition estimate for Network-Friend, Network-Temporal, and Temporal-Friend, pairwise classifiers from Botometer, for the 35,308 Twitter accounts in our first dataset, obtained through our BoostNet method. The regions in the upper right corners correspond to the over 3,000 socialbot accounts that we discovered. These results were obtained on April 9–18, 2018

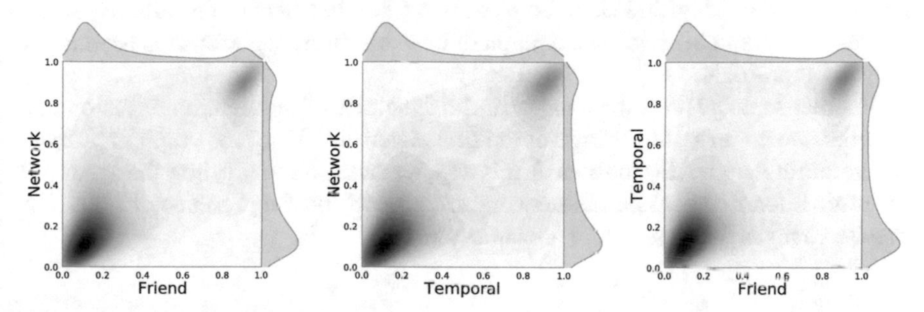

Fig. 4 2D Kernel decomposition estimate for Network-Friend, Network-Temporal, and Temporal-Friend, pairwise classifiers from Botometer, for the 12,044 Twitter accounts in our second dataset, obtained through our BoostNet method. The regions in the upper right corners correspond to the over 2154 socialbot accounts that we discovered. These results were obtained between November 2018 and January 2019

Table 1 Linked accounts in our two datasets, from the full networks BoostNet reconstructed in April and in November 2018. A comparison between both datasets yields 3688 shared accounts, and at least 646 of them were classified as socialbots

Accounts in network		
	Total	Bots
April	35,208	3009
November	12,044	2154

kinds of automated accounts (when they are present) and also provides an insight into the limitation of choosing a pre-established threshold to decide whether an account should be considered a bot, or not.

A numerical summary of the number of accounts found appears in Table 1.

5 Topological Network Effect of Bots

How can we quantify the topological effects of bot accounts in a network? The third named author together with Leo Torres and Tina Eliassi-Rad defined a new graph distance that utilizes topological information from the fundamental group (first homotopy group) of the underlying graph, and metric information in its marked length spectrum [16]. In order for this method to be applied here, we consider each of the two datasets (from April and from November, both in 2018) and we add a single node to each one of the associated follower–followee graphs. In this way, we now have connected graphs G'_A and G'_N. From each of these graphs, we now delete the nodes that have been classified as bots, and obtain two graphs G_A and G_N. Now we compute the spectral non-backtracking distances $d(G_A, G'_A)$ and $d(G_N, G'_N)$, which provide a topologically informed way to measure the network effects of bots. The results are included in Table 2. In both cases, we see that there is a positive distance. However, a comprehensive investigation of how this distance performs is beyond the scope of this article.

In order to understand the change in the dynamics of information diffusion in a network, we compare the change of the first eigenvalue λ_G of the graph Laplacian, before and after removing the bots. In this way, we can also quantify how the dynamics of information flows inside the network are affected by the presence of bots. The precise changes for these values are shown in Table 3.

Table 2 Measurement of topological changes in the two datasets, as the bot accounts are removed and the spectral non-backtracking distance (using Wasserstein distance between eigenvalues) is computed between the two resulting networks. For more on this method and its validation with random graph models as well as empirical data, we refer interested readers to [16]

Topological effect of bots in a network	
	Non-backtracking distance
April	$d(G_A, G'_A) = 0.0997397511933$
November	$d(G_N, G'_N) = 0.0856553118609$

Table 3 An approximation to information diffusion changes in the two datasets can be measured using the changes in the first eigenvalue of the graph Laplacian

Information diffusion effect of bots in a network	
	Change in Laplacian 1st eigenvalue
April	$\lambda_{G_A} - \lambda_{G'_A} = 1504.5$
November	$\lambda_{G_N} - \lambda_{G'_N} = 330.0$

6 Conclusions

Our work here demonstrates how statistical methods can show the existence of considerable socialbot network of linked accounts. Given the potential size of Guatemala's total Twitter user base, the amount of socialbot accounts could certainly impede freedom of expression. These findings corroborate the experience of users (and journalists) who claimed widespread abuse of this technology for nefarious purposes was present in Guatemala.

Moreover, our BoostNet strategy can be employed in other circumstances and social media platforms, where limited observational data can then lead to a complete reconstruction of networks of malicious accounts.

We hope that various statistical, topological, and dynamical approaches used here will be of interest to researchers and social media users who face these kinds of circumstances.

Acknowledgements We thank the OSoMe team in Indiana University for access to *Botometer* , and also Twitter for allowing access to data through their APIs. PSS acknowledges support from UNAM-DGAPA-PAPIIT-IN104819 and thanks IPAM, UCLA for an excellent and stimulating environment during the final stages of this work.

References

1. Aral, S., Dhillon, P.S.: Social influence maximization under empirical influence models. Nat. Hum. Behav. **1** (2018)
2. Chavoshi, N., Hamooni, H., Mueen, A.: Identifying correlated bots in twitter. In: International Conference on Social Informatics, pp. 14–21. Springer International Publishing, New York (2016)
3. Chu, Z., Gianvecchio, S., Wang, H., Jajodia, S.: Who is tweeting on twitter: human, bot, or cyborg? In: Proceedings of the 26th Annual Computer Security Applications Conference, ACSAC'10, pp. 21–30. ACM, New York (2010)
4. CICIG (International Commission against Impunity in Guatemala), United Nations. U.N. Department of Political Affairs. https://dpa.un.org/en/mission/cicig. Accessed 28 Dec 2018
5. Clark, E.M., Williams, J.R., Jones, C.A., Galbraith, R.A., Danforth, C.M., Dodds, P.S.: Sifting robotic from organic text: a natural language approach for detecting automation on twitter. J. Comput. Sci. **16**, 1–7 (2016)
6. Davis, C.A., Varol, O., Ferrara, E., Flammini, A., Menczer, F.: BotOrNot: A system to evaluate social bots. In: Proceedings of the 25th International Conference Companion on World Wide Web, WWW'16 Companion, pp. 273–274. International World Wide Web Conferences Steering Committee (2016)
7. Dickerson, J.P., Kagan, V., Subrahmanian, V.: Using sentiment to detect bots on twitter: are humans more opinionated than bots? In: 2014 IEEE/ACM International Conference on Advances in Social Networks Analysis and Mining (ASONAM) 00(undefined), pp. 620–627 (2014)
8. Ferrara, E., Varol, O., Davis, C., Menczer, F., Flammini, A.: The rise of social bots. Commun. ACM **59**(7), 96–104 (2016)

9. Gallagher, E., Suárez-Serrato, P., Velazquez Richards, E.I.: Socialbots whitewashing contested elections; a case study from honduras. In: Third International Congress on Information and Communication Technology, Advances in Intelligent Systems and Computing, vol. 797, pp. 547–552. Springer, Singapore (2019)
10. Internet Live Stats. http://www.internetlivestats.com/internet-users/guatemala/. Accessed 11 Jan 2019
11. Linthicum, K.: A U.N. anti-corruption commission is fleeing Guatemala after president's order, Los Angeles Times, Mexico and The Americas, 8 Jan 2019. https://lat.ms/2M4HB2S
12. Pew Research Center: U.S. unauthorized immigrant total dips to lowest level in a decade (2018). https://pewrsr.ch/2Qptbid. Accessed 26 Dec 2018
13. Social Media Stats Guatemala, GlobalStats. http://gs.statcounter.com/social-media-stats/all/guatemala. Accessed 11 Jan 2019
14. Stella, M., Ferrara, E., De Domenico, M.: Bots increase exposure to negative and inflammatory content in online social systems. Proc. Natl. Acad. Sci. **115**(49), 12435–12440 (2018)
15. Suárez-Serrato, P., Roberts, M.E., Davis, C., Menczer, F.: On the influence of social bots in online protests. In: Spiro, E., Ahn, Y.Y. (eds.) Social Informatics. SocInfo 2016. Lecture Notes in Computer Science, vol. 10047. Springer, Cham (2016)
16. Torres, L., Suárez-Serrato, P., Eliassi-Rad, T.: Non-backtracking cycles: length spectrum theory and graph mining applications. Appli. Netw. Sci. **4**(1), 41 (2019). https://doi.org/10.1007/s41109-019-0147-y
17. Varol, O., Ferrara, E., Davis, C.A., Menczer, F., Flammini, A.: Online human-bot interactions: detection, estimation, and characterization. In: Proceedings of the Eleventh International Conference on Web and Social Media, ICWSM 2017, Montréal, Québec, Canada, 15–18 May 2017, pp. 280–289
18. Varol, O., Ferrara, E., Menczer, F., Flammini, A.: Early detection of promoted campaigns on social media, EPJ data science, pp. 2193–1127 (2017)
19. Velázquez, E., Yazdani, M., Suárez-Serrato, P.: Socialbots supporting human rights. In: Proceedings of the 2018 AAAI/ACM Conference on AI, Ethics, and Society, AIES 2018, New Orleans, LA, USA, 02–03 Feb 2018, pp. 290–296 (2018). https://doi.org/10.1145/3278721.3278734
20. Woolley, S.: Automating power: social bot interference in global politics. First Monday **21**(4) (2016)

Printed in the United States
By Bookmasters